The Institute of Biology's
Studies in Biology no. 4

An Introduction to Parasitology

Second Edition

R. Alan Wilson
Ph.D., D.C.C., D.I.C.
Senior Lecturer, Department of Biology,
University of York

Edward Arnold

© R. Alan Wilson, 1979

First published 1967
by Edward Arnold (Publishers) Limited
41 Bedford Square, London WC1B 3DQ

Reprinted 1969
Reprinted 1971
Reprinted 1974
Second Edition 1979
Reprinted 1983

Paper edition ISBN: 0 7131 2750 3

British Library Cataloguing in Publication Data

Wilson, Robert Alan
  An introduction to parasitology. – 2nd ed – (Institute of Biology.
  Studies in biology; No. 4: 0537–9024).
  1. Parasitology
  I. Title   II. Series
  574.5'24   QL757

  ISBN 0–7131–2750–3

Printed and bound in Great Britain at
The Camelot Press Ltd, Southampton

General Preface to the Series

Because it is no longer possible for one textbook to cover the whole field of biology while remaining sufficiently up to date, the Institute of Biology proposed this series so that teachers and students can learn about significant developments. The enthusiastic acceptance of 'Studies in Biology' shows that the books are providing authoritative views of biological topics.

The features of the series include the attention given to methods, the selected list of books for further reading and, wherever possible, suggestions for practical work.

Readers' comments will be welcomed by the Education Officer of the Institute.

1983

Institute of Biology
20 Queensbury Place
London SW7 2DZ

Preface to the Second Edition

In this short book I have given a personal view of those aspects of parasitology in which there is currently a lively research interest. Since the publication of the first edition there have been some notable shifts of emphasis within parasitology. Studies on the immunology of parasite infections have increased, with the production of anti-parasite vaccines as their ultimate, if elusive, goal. In parasite epidemiology there is now a concerted effort to place the understanding of transmission dynamics on a quantitative base.

In the choice of topics and the examples with which they are illustrated, I have necessarily been selective. There is a bias towards parasites of medical or veterinary importance with a consequent neglect of countless other interesting associations.

Animal parasitology has recently received a considerable impetus from the designation by the World Health Organisation of six major tropical diseases as targets for research. Five of these, malaria, trypanosomiasis, leishmaniasis, schistosomiasis and filariasis fall within the scope of this book.

I would like to thank the authors whose illustrations are acknowledged in the various figures. I am also indebted to my wife, Margaret, for checking the text, and to many others who have assisted in various ways.

York, 1978
R.A.W.

Contents

1 What is a Parasite?

Parasitism is just one among many types of association between two organisms and there is no single feature which can be used to label an animal indubitably as a parasite. We must therefore examine the phenomenon from several viewpoints if we are to form a coherent picture.

The parasite obtains food at the expense of its host by consuming either host tissues and fluids, or the contents of the host intestine. The relationship of parasite to host therefore has a nutritional basis. How then do we distinguish between parasitism and other ways of acquiring food, such as predation or scavenging? An obvious answer is to regard parasitism as a special form of predation in which the host is not killed in providing the parasite with a meal. However, many small animals such as blood-sucking insects feed on larger animals without killing them. Are mosquitoes or fleas to be considered as parasites or predators? The criterion for deciding is clearly the length of time which the feeding animal remains with the host. Most parasitologists would agree that the stay must be a significant length of time, certainly more than the few seconds for which a biting fly feeds: they would disagree about the minimum duration of stay needed to qualify the feeder as a parasite.

It is necessary to distinguish still further between parasitism and other nutritional associations such as commensalism and symbiosis, in which the associates remain together for long periods. If the association results in mutual benefit to the participants then clearly it cannot be classed as parasitism. Conversely, if one of the associates harms the other then we must consider it to be a parasite. The problem is that it has proved impossible to demonstrate that many so-called parasites harm the host in any way.

Finally, we need to consider a unique property of the environment provided for the parasite by the host. The parasite may be recognized as a foreign invader against which the host can mount an immune response. A measure of a parasite's success is its ability to evade this response which is aimed at its elimination.

The animals dealt with in this book almost invariably cause disease in their hosts and in practice, the reader will have no difficulty in identifying them unequivocally as parasites. The fact that a parasite causes disease has often been considered a measure of its poor adaptation, but that is to take an oversimplified view of the nature of parasitism. It is now apparent that some parasites are important natural regulators of host populations. The excellent adaptation of parasites can equally well be measured by the limited success of schemes aimed at their control and eradication.

2 Life Cycles

2.1 The scope of parasitology

In the broadest sense, parasitology is the study of those organisms which spend all or part of their lives as parasites in or on other organisms. In practice it has a more restricted usage principally covering the study of parasitic Protozoa, Platyhelminthes and Nematoda. These three groups contain the majority of parasites of medical and veterinary importance which form the subject matter of this book. Other invertebrate groups with significant parasites include acanthocephalans, leeches, ticks, crustacea and insects. For a wider view of these the reader should consult the general references listed at the end of the book.

From another standpoint, parasitology can be viewed as a microcosm of biology embracing subjects as disparate as molecular biochemisty and the mathematics of population processes. In this chapter the reader is introduced to the parasites with their intricate life cycles. Subsequent chapters deal with some of the major areas on which research interest is currently focused.

2.2 Protozoa

The phylum Protozoa contains several thousand parasitic species. Members of two groups, the Sarcomastigophora (amoebae and flagellates) and the Sporozoa (coccidia and malaria) are described here. The parasitic Protozoa are unicellular and some are so small that their structure is hard to resolve by light microscopy. However, the application of electron microscopy has revealed that these minute scraps of protoplasm have a complex and hitherto unsuspected ultrastructure.

2.2.1 Parasitic amoebae

A number of genera of amoebae are parasitic in various parts of the alimentary tract of vertebrates and invertebrates. There are several non-pathogenic species in man together with one important pathogen *Entamoeba histolytica*, found in the colon. The trophozoite, the feeding stage, has a typical amoeboid form using pseudopodia for movement and phagocytosis. In the lumen of the colon it feeds on bacteria and cell debris, multiplying by binary fission. For unknown reasons, these lumen-dwelling trophozoites may become pathogenic. They invade the mucosa of the colon, feed on blood and tissues, and produce ulcers. In a proportion of infections, trophozoites are carried in hepatic portal blood to the liver and become established there forming liver abscesses. These abscesses may enlarge due to the activities of amoebae at the periphery, destroying liver tissue. From the liver the amoebae can spread to adjacent organs, particularly the lungs and pleural cavity.

The trophozoites dwelling in the lumen of the intestine are capable of forming resistant cysts which pass out with the faeces. Infection of another host results from ingestion of these cysts.

2.2.2   Parasitic flagellates

There are two main groups of parasitic flagellates: one group is found in the alimentary and genital tracts, and the other in the tissues and blood stream of vertebrates. The lumen-dwellers include *Trichomonas*, *Giardia* and *Histomonas*. Members of the genus *Trichomonas* have simple life cycles with transmission of the unprotected trophozoite by direct contact. *Giardia* causes a form of dysentery in man and, like parasitic amoebae, is transmitted by resistant cysts. *Histomonas*, parasitic in the caecum of turkeys and other gallinaceous birds, has an unusual mode of transmission. The trophozoites invade the tissues of another gut parasite, the nematode *Heterakis* and are passed on to a new host in its eggs.

The tissue and blood-dwelling flagellates are often referred to as haemoflagellates. Because the site of infection is enclosed, transmission to a new vertebrate host is by means of an intermediate host or vector, usually a blood-feeding insect. The haemoflagellates are thought to have evolved from genera similar to *Leptomonas*, parasitic in the intestine of insects. The trophozoites of the haemoflagellates are polymorphic, the different forms being termed amastigote, promastigote, epimastigote and trypomastigote according to the positions of the kinetoplast (a prominent cellular feature) and flagellum, relative to the nucleus (see Figs 2–1 and 2–2). The trypomastigote of *Trypanosoma* (Fig. 2–1) has an elongate body to which a single flagellum is attached by an undulating membrane, and it swims through the blood with a rapid jerking motion. The flagellum inserts at one end of the cell near the kinetoplast, a specialized region of the single elongate mitochondrion, distinguished by the presence of cytoplasmic DNA. The trypomastigote has a cytostome which may ingest material by pinocytosis. The exterior of the cell is covered by a glycoprotein coat. It is unlikely that sexual reproduction occurs in haemoflagellates and multiplication is by binary fission.

Members of the genus *Trypanosoma* are important parasites of man and his livestock. There are two groups divided according to their mode of transmission by the insect vector: the stercoraria are transmitted via the faeces and the salivaria via the salivary glands.

The life cycle of *T. cruzi*, a parasite of man in South and Central America causing Chagas' disease, is illustrated in Fig. 2–2. Chagas' disease is a zoonosis, i.e. a disease with a large number of wild mammals

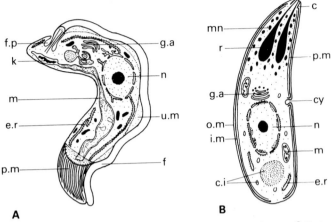

Fig. 2-1 Body form of protozoans. (A) The trypomastigote of *Trypanosoma congolense* (after VICKERMAN K., (1969). *J. Protozool.*, **16**, 54–69). (B) A generalized sporozoite/merozoite, typical of the genera *Eimeria* and *Plasmodium*. c – conoid, c.i – cytoplasmic inclusions, cy – cytosome, e.r – endoplasmic reticulum, f – flagellum, f.p – flagellar pocket, g.a – golgi apparatus, i.m – inner membrane, k – kinetoplast, m – mitochondrion, mn – micronemes, n – nucleus, o. m – outer membrane, p.m – pellicular microtubules, r – rhoptries, u. m – undulating membrane.

acting as reservoir hosts. The vectors are blood-feeding bugs. Infection of man results when faeces containing metacyclic (between cycles) forms contaminate the bite wound or are transferred to the mucous membranes of the mouth. Multiplication of the amastigote results in the appearance of sac-like pseudocysts, particularly in cardiac muscle. The pseudocysts burst, releasing trypomastigotes into the bloodstream. Reinvasion of tissue or ingestion by the vector may occur. There is further multiplication in the vector and the infective metacyclic forms develop in its hind-gut.

There are numerous pathogenic species of salivarian trypanosomes in Africa. *T. brucei*, *T. vivax* and *T. congolense* attack cattle. *T. rhodesiense* and *T. gambiense* are the cause of sleeping sickness in man. (*T. rhodesiense* is another example of a widespread zoonosis.) The insect intermediate hosts are tsetse flies of the genus *Glossina*. Metacyclic trypanosomes are found in the salivary glands of the fly, and infection of man results from the direct inoculation of these forms into the skin when it feeds. In the bloodstream, the trypomastigotes develop into three forms: slender, intermediate and stumpy. The slender forms multiply most frequently; the stumpy are infective to the tsetse fly host. In the fly mid-gut the trypomastigotes multiply and eventually migrate forward to enter the salivary glands. Here they assume the epimastigote form and undergo further multiplication giving rise to the infective metacyclic trypanosomes.

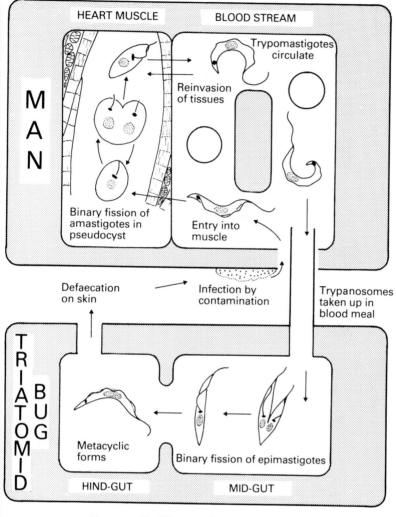

Fig. 2-2 The life cycle of *Trypanosoma cruzi*.

Members of the genus *Leishmania* are another group of haemoflagellates which parasitize man. They are transmitted by female sandflies of the genera *Phlebotomus* and *Lutzomyia*. The amastigote form invades macrophages of the reticulo-endothelial system of man. The amastigotes multiply and are released when the cells burst. They may then invade neighbouring macrophages producing a spreading lesion (see Chapter 7). The sandfly becomes infected when it takes a blood meal and multiplication of the parasite in the promastigote form occurs in the mid-gut. From here the promastigotes pass to the pharynx and proboscis and are inoculated into a new host when the fly takes its next meal.

The different leishmanias are distinguished by their antigenic structure, rather than by morphological criteria, and several species groups are recognized. *L. tropica* occurs principally in the Middle East and causes a disease of the skin called Oriental Sore. *L. braziliensis* occurs in South and Central America and attacks the skin and the mucous membranes of the mouth, nose, etc. *L. donovani* occurs in Southern Europe, Africa and Asia and causes visceral leishmaniasis or Kala-azar. In this disease the parasite spreads rapidly to the macrophages of the spleen, liver and other internal organs. All leishmaniases are zoonoses with a range of mammal reservoir hosts.

2.1.3 Sporozoa

The Sporozoa are a numerous and exclusively parasitic group of Protozoa. They are thought to have originated as extracellular intestinal parasites of invertebrates, transmitted by resistant cysts. Some species presumably became parasitic in intestinal tissues, but retained transmission via cysts in the faeces. Eventually parasites of the bloodstream evolved and acquired insect vectors.

The Coccidia are intracellular parasites of the alimentary tract. Species of the genus *Eimeria* are responsible for the disease coccidiosis in poultry, and also infect other farm livestock. The life cycle of *Eimeria* is direct with transmission by means of sporozoites (Fig. 2–1) enclosed within a protective oocyst. The sporozoites are released into the lumen of the intestine when the oocyst is ingested by the host. They penetrate single epithelial cells and multiply by a process of schizogony in which the nucleus of the parasite divides several times but its cytoplasm remains intact. The daughter nuclei migrate into outgrowths at the periphery of the schizont and differentiation into merozoites occurs. These escape to the gut lumen and invade further epithelial cells to repeat the schizogony. In this manner, the entire intestinal epithelium may be destroyed, resulting in death of the host. Schizogony is followed by sexual reproduction. merozoites enter intestinal cells and develop into either microgametocytes or macrogametocytes. Motile microgametes are formed and escape into the intestinal lumen. They then fuse with the much larger macrogametes to form a zygote. The zygote secretes a protective coat to become the oocyst and is passed out in the faeces. The final phase of development, sporogony, takes place outside the host, resulting in the formation of eight sporozoites within the oocyst.

Another widely distributed protozoan, *Toxoplasma gondii*, is now recognized as a coccidian. The trophozoites, contained in cysts up to 0.1 mm in diameter, have been identified in the tissues of many mammals and birds. In some regions of the world up to 90% of the human population may be infected. The infection is usually asymptomatic but in situations where resistance is lowered, the trophozoites may proliferate rapidly, causing death. *Toxoplasma* can also be transmitted congenitally to the foetus if contracted during early pregnancy and the foetus may be damaged or aborted. In 1969 it was demonstrated that *Toxoplasma* can behave like a typical coccidian with cycles of schizogony and gamete formation in the small intestine of the cat. Oocysts are passed in the faeces, and when sporulated are remarkably similar to those of the coccidian genus *Isospora*. Two routes of infection are therefore possible in man: ingestion of the tissue cysts containing trophozoites, or ingestion of oocysts acquired from cat faeces.

The blood-dwelling Sporozoa include the genus *Plasmodium*, four species of which (*P. vivax*, *P. ovale*, *P. malariae* and *P. falciparum*) cause malaria in man. Although eradication programmes have been undertaken on a huge scale for more than twenty years, malaria is still arguably the most important transmissible human disease. The vectors are female mosquitoes of the genus *Anopheles* and the life cycle of *P. falciparum* is illustrated in Fig. 2–3. Man becomes infected when sporozoites are inoculated into the bloodstream in the saliva of a feeding mosquito. The first cycle of schizogony takes place in the liver parenchyma. The resulting merozoites then enter erythrocytes and further cycles of schizogony occur, producing a rapid rise in parasitaemia (the number of parasites detectable in the blood). Eventually gametogony takes place in the erythrocytes and the circulating gametocytes are taken up by the mosquito when it feeds. The gametes develop rapidly in the mid-gut of the mosquito. The motile microgamete locates and fuses with a macrogamete to form a zygote, which then penetrates the mid-gut epithelium. Sporogony takes place in large oocysts which lie on the outside of the mid-gut wall. Sporozoites infective to man are released into the body cavity when these oocysts rupture, and eventually enter the insect's salivary glands.

Two other blood-dwelling haemosporidia which infect cattle, deserve a mention. These are *Babesia bigemina* with worldwide distribution causing red-water fever, and *Theileria parva* in East Africa causing East Coast fever. Both parasites are transmitted by cattle ticks.

2.3   Platyhelminthes

There are three major groups of parasitic flatworms, the Monogenea, Digenea and Cestoda, whose interrelationships are not clear. The Monogenea were formerly classed with the Digenea as the Trematoda but some authorities now consider that their closest affinities are with the Cestoda. Both monogeneans and digeneans possess a functional alimentary tract, but cestodes have no trace of this organ at any stage in their life cycle. The body surface of a parasitic flatworm (Fig. 2–4) was originally described as being covered with an inert cuticle. It is in fact a

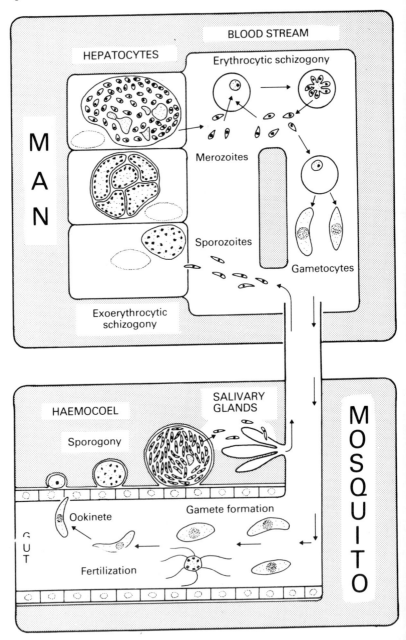

Fig. 2–3 The life cycle of *Plasmodium falciparum*.

syncytial layer of cytoplasm (i.e. without transverse walls) bounded on inner and outer surfaces by a plasmamembrane. It is connected by numerous narrow cytoplasmic tubules to nucleated cell bodies which lie beneath the musculature of the body wall. The layer is termed a tegument to distinguish it from an inert and secreted cuticle such as that of nematodes. The parasitic flatworms are generally hermaphrodite.

The body organization of a typical adult digenean is illustrated in Fig. 2–4. There are generally prominent oral and ventral suckers for

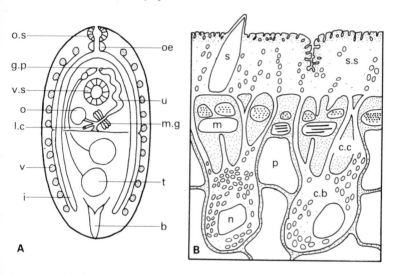

Fig. 2–4   Body form of digenetic flatworms. (**A**) Diagrammatic plan of the worm. (**B**) Section through the tegument of *Fasciola hepatica* (after THREADGOLD, L. T., (1963). *Quarterly J. of micr. Sci.*, 104, 502–12). b – bladder, c.b – cell body, c.c – cytoplasmic connections, g.p – genital pore, i – intestine, l.c – Laurer's canal, m – muscle, m.g – Mehlis' gland, n – nucleus, o – ovary, oe – oesophagus, o.s – oral sucker, p – parenchyma, s – spine, s.s – surface syncytium, t – testis, u – uterus, v – vitellaria, v.s – ventral sucker.

attachment to the host tissues. Internal organs include a bifurcated gut, a protonephridial system and separate male and female genitalia. The nervous system is relatively simple, consisting of paired anterior ganglia with connections to the musculature and peripheral sense organs.

2.3.1   Monogenea

These flatworms are chiefly ectoparasites of cold-blooded aquatic vertebrates. The organization of the body is in most aspects similar to that of the Digenea. However, monogeneans attach to the host using a specialized posterior opisthaptor. In skin parasites it takes the form of a simple sucker-like disc with associated hooks. In gill-dwellers there may be an arrangement of suckers and clamps for attachment to the gill filaments.

The life cycle of monogeneans is direct. The mature worm lays shelled, operculate eggs which are either attached to host tissues or released into the water. A minute ciliated larva, the oncomiracidium, develops and hatches from the egg. If the free-swimming larva locates a host it will attach and grow directly into an adult worm.

Monogenea are not generally pathogenic but in circumstances where host population densities are artificially increased (e.g. fish hatcheries or fish farms) the normal balance between host and parasite populations may be altered. This leads to a much increased parasite burden and mortality in the host fish population.

2.3.2   Digenea

The Digenea have complex life cycles involving two or three hosts and the adult worms parasitize virtually all the organ systems of vertebrates. The first intermediate host is, with few exceptions, a gastropod mollusc. When there is a second intermediate host in the cycle it often features in the diet of the vertebrate final host. Transmission is generally effected by two free-swimming larval stages, and most digenean life cycles are therefore dependant upon water for their completion.

Members of the genus *Schistosoma* are important parasites of man and domestic livestock. The three species which infect man are *S. mansoni* in Central and South America, and Africa; *S. haematobium* in Africa and the Middle East; *S. japonicum* (a zoonosis) in China, Japan and the Philippines. Collectively these three species are second only to malaria in medical importance, with an estimated 150–200 million people currently infected. The schistosomes are unusual in having separate sexes. The adult worms are thread-like, about 1 cm in length, and inhabit the blood vessels of the hepatic portal system. The life cycle of *S. mansoni* is illustrated in Fig. 2–5.

The female worm deposits undeveloped spined eggs in the tissues of the intestine. There the eggs develop and by abrasion of the tissues reach the lumen. They pass out with the faeces and hatch if they enter freshwater. The miracidium is a non-feeding larval stage capable of rapid swimming. It either locates and penetrates through the epidermis of a suitable snail host (various species of the genus *Biomphalaria*) or dies within 24 hours of hatching. Immediately after entering the snail's haemocoel, the miracidium metamorphoses into the sac-like mother sporocyst. This is a feeding stage but, lacking a gut, must acquire nutrients by diffusion or active transport. Within the body of the sporocyst the second generation, consisting of numerous daughter sporocysts, develops by asexual multiplication. The hepato-pancreas of the snail becomes packed with daughter sporocysts, inside which a further phase of asexual multiplication gives rise to the third generation, the cercaria larvae. When

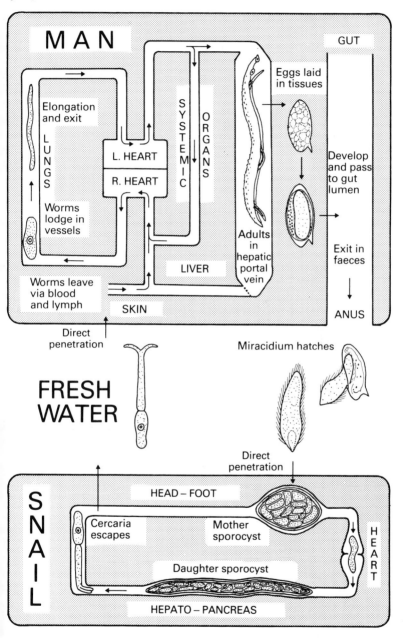

Fig. 2–5 Life cycle of *Schistosoma mansoni*.

these are mature, they leave the daughter sporocyst and migrate via the haemocoel of the snail to the mantle cavity where they escape into freshwater. The cercaria, which possesses a forked tail used for propulsion, is also a non-feeding larva with a brief life.

Man becomes infected by contact with water containing cercariae. The larvae possess special gland cells, the secretions of which enable them to penetrate human skin directly. Within hours of entering the skin, the cercaria completes a metamorphosis into the schistosomulum stage. It remains there for at least 48 hours before migrating to the hepatic portal system. Several routes have been postulated and one, the intravascular route, is illustrated in Fig. 2–5. A schistosomulum might need to make several circuits of the vascular system before arriving by chance at the hepatic portal system. It is then inhibited from further migration and stimulated to mature. After about three weeks the male and female worms pair. The male worm clasps the female in a ventral groove, the gynaecophoric canal, and carries her against the flow of blood to the mesenteric capillaries where she commences to lay eggs.

Because the schistosome life cycle is intimately linked with snails and water, the disease is especially prevalent in areas where irrigation is practised (e.g. the Nile Valley).

The common liver fluke *Fasciola hepatica* parasitizes domestic livestock, and occasionally man, with a world-wide distribution. The leaf-shaped mature worms, up to 3 cm long, live in the bile ducts, feeding on tissue and blood. Eggs pass out with the faeces and develop in freshwater. The free-swimming miracidium infects the mud snail *Lymnaea truncatula* (and other lymnaeids) and metamorphoses into a sporocyst. The second generation larva within the snail is termed a redia. It possesses a sac-like gut and consumes the gonads and hepatopancreas. The redial generation gives rise either to daughter rediae, or to cercariae which exit from the snail. After a brief swimming period the cercaria attaches to vegetation and secretes a resistant cyst. It is now a metacercaria infective to livestock which ingest it when grazing. When the cyst reaches the small intestine, the larva escapes, bores through the intestinal wall and makes its way to the surface of the liver. It then eats its way through the liver tissue to reach the bile ducts. This is the period when the host is most at risk, and death can result from damage caused by the invasion. The fluke also creates the conditions in which the bacterium *Clostridium oedimatiens* infects the liver, causing Black's disease.

Fasciola is particularly common in regions with high rainfall or poor drainage where the prevailing wet conditions favour both the snail and transmission of the parasite.

Dicrocoelium dendriticum is another important liver fluke of grazing animals. In Britain it is restricted to the Hebrides and land snails such as *Helicella itala* serve as the first intermediate host. Livestock become infected when they accidentally ingest metacercariae contained within the body of the second intermediate host, an ant of the genus *Formica*.

Three other digenean parasites of man in Eastern Asia deserve a mention. *Paragonimus westermani* infects the lungs and pleura. Its cercariae encyst in various species of freshwater crabs and crayfish. *Clonorchis sinensis* infects the bile ducts of the liver and there are an estimated 30 million cases in man. The cercaria encysts in the tissues of a variety of freshwater fish. *Fasciolopsis buski* parasitizes the small intestine and can grow up to 7 cm in length. Its life cycle is similar to that of *Fasciola hepatica* and the cercaria encysts on freshwater plants, such as the water chestnut, which are cultivated for food.

2.3.3   Cestoda

The Cestoda, or tapeworms, are parasites chiefly of the alimentary tract of vertebrates. They have complex life cycles with one or two intermediate hosts and almost invariably gain entry to the vertebrate final host via its food. The mature tapeworm attaches to the wall of the intestine using devices such as hooks or suckers borne on the scolex (head). Immediately behind the scolex is the neck, a region of cell division and differentiation, where the proglottids (segments) are generated. Each contains a complete set of reproductive organs which mature as it is progressively displaced towards the posterior. The chain of proglottids which can be up to several metres in length, is called the strobila. This replication of the facilities for egg production should be viewed as the tapeworm's solution to the problems of propagation of the species posed by a parasitic way of life.

A great deal of research has been carried out on *Hymenolepis diminuta*, the rat tapeworm, because of the ease with which it can be maintained in the laboratory. Its life cycle is illustrated in Fig. 2–6. Mature proglottids detach from the end of the strobila of the adult worm in the small intestine and pass out in the faeces. A number of insects can serve as intermediate hosts and in the laboratory, the flour beetle *Tribolium confusum* is used. The egg contains a six-hooked larva, the hexacanth, which can remain infective for long periods and is released into the gut lumen after ingestion by the beetle. The hexacanth uses a combination of hook movements and lytic secretions to gain access to the haemocoel of the beetle. There it grows and differentiates into the cysticercoid larva. The cysticercoid is a complex structure consisting of a minute scolex and neck region surrounded by a series of protective layers. When the beetle is swallowed by a rat, the scolex is liberated into the small intestine by host digestive enzymes. It attaches to the gut wall and grows to maturity.

The life cycles of *Taenia solium* and *Taenia saginata*, which parasitize man, are very similar to that of *H. diminuta*. However, the intermediate hosts are pigs and cattle, respectively, and the infective larval stage is a bladderworm or cysticercus. This is found in the muscles and man becomes infected by eating raw meat containing cysticerci, in such dishes as Steak Tartare. The main effects of harbouring a mature tapeworm

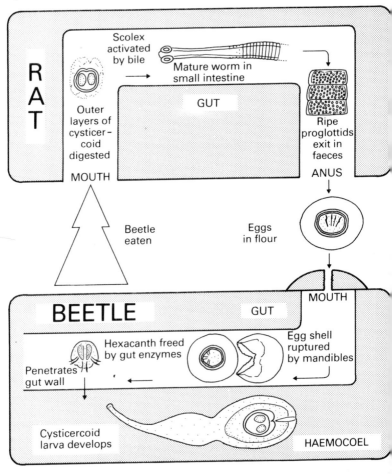

Fig. 2–6 Life cycle of *Hymenolepis diminuta*.

result from its competition with the host for the available food, leading to undernourishment. The larval stages cause more severe problems. The eggs of *Taenia solium* are infective to man and can on rare occasions cause the disease, cysticercosis, in which bladder worms become distributed throughout the body. If they are deposited in the brain nervous disorders may be generated.

The most widespread and dangerous form of cysticercosis is known as hydatid disease. Large hydatid cysts, containing the larvae of the dog tapeworm *Echinococcus* may develop in man. The normal intermediate host is the sheep but man becomes infected by acquiring eggs from dog faeces. The hexacanth larvae bore through the wall of the small intestine and are carried in the hepatic portal blood to the liver from which a few pass to other organs such as the lungs. The hydatid cyst is capable of budding daughter cysts from its internal surface and many scolesces are formed. In time, a hydatid cyst may attain a volume of several litres. The host's immune system becomes sensitized to the cyst and death from anaphylactic shock results if it ruptures either spontaneously or during surgical removal.

The broad tapeworm of man, *Diphyllobothrium latum*, has a more typical mode of transmission. The eggs, remarkably similar to those of digeneans such as *Fasciola*, embryonate in freshwater and a free-swimming non-feeding coracidium larva hatches. The first intermediate host is a copepod crustacean in which a procercoid larva develops. Plerocercoid larvae, infective to man, develop in the tissues of fish which feed on the copepods.

As the life cycle of *D. latum* is intimately connected with freshwater, the tapeworm is commonest in countries with extensive lake systems, such as Finland or Switzerland. *D. latum* is able to accumulate vast amounts of vitamin B_{12} in its tissues and as a consequence may deprive the human host of this factor, causing pernicious anaemia.

2.4 Nematoda

The Nematoda are a widespread and successful group found in marine, freshwater and terrestrial habitats as well as parasitizing plants and animals. The sexes are usually separate but in some species where males have not been found, reproduction is parthenogenetic. A diagrammatic section through the body wall is illustrated in Fig. 2–7. The wall is bounded on the outer surface by a cuticle secreted by an underlying syncytial hypodermis. Beneath this are four bands of longitudinal muscle (one is shown in the diagram). In locomotion, the contraction of these muscles is opposed by the turgor of the fluid-filled body cavity and a meshwork of fibrils in the cuticle. The alimentary tract and gonads lie in the body cavity. The gonads take the form of single or paired coiled tubes. A number of accessory structures such as spicules and caudal alae may be present in the male to aid in copulation.

Nematodes are either oviparous or ovoviviparous (forming eggs which develop and hatch whilst still in the body of the female). The larvae are similar in appearance to adult worms but lack gonads. Development into the adult is punctuated by a series of moults, almost always four in number. At each moult a new cuticle is secreted by the underlying hypodermis and the old one is shed.

The classification of nematodes is beset with many problems. Not least of these is the remarkable similarity of body plan throughout the group making precise identification difficult, particularly of larval stages. Several groups of nematodes appear to have independently evolved a

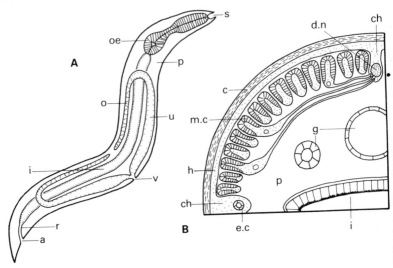

Fig. 2–7 Body form of nematodes. (A) Diagrammatic plan of a female nematode. (B) Transverse section through the body of *Ascaris lumbricoides* (one quarter only). a – anus, c – cuticle, ch – chord, d.n – dorsal nerve, e.c – excretory canal (lateral), g – gonad, h – hypodermis, i – intestine, m.c – muscle cell, oe – oesophagus (pharynx), o – ovary, p – pseudocoelom, r – rectum, s – stoma, u – uterus, v – vulva.

parasitic way of life. Examples of four types, selected for their medical or veterinary importance, are discussed below.

2.4.1 Strongylida

This group includes the hookworms of man and the gastro-intestinal nematodes of grazing animals. The life cycle of *Ancylostoma duodenale*, which is direct, is illustrated in Fig. 2–8. This nematode, together with *Necator americanus*, is common throughout the tropics and subtropics with an estimated 700 million human cases. The mature worms are about 1 cm long and attach to the villi of the small intestine by means of chitinous teeth present in the buccal cavity. Thin-shelled oval eggs are laid and pass out with the faeces. If they reach water or moist soil, development occurs and a free-living, first-stage larva hatches. It feeds on faecal material and moults twice to become the third larval stage. The second moult is incomplete, a new cuticle being secreted, although the old is not shed. This ensheathed larva is a non-feeding stage, infective to man, and can remain viable for some weeks, depending on environmental conditions. If the third-stage larva contacts human skin, usually the feet or legs, it is stimulated to penetrate directly. It enters the bloodstream and migrates via the heart to the lungs. There, it breaks out of the capillaries into the alveoli and moults again to become the fourth stage. It is carried up the

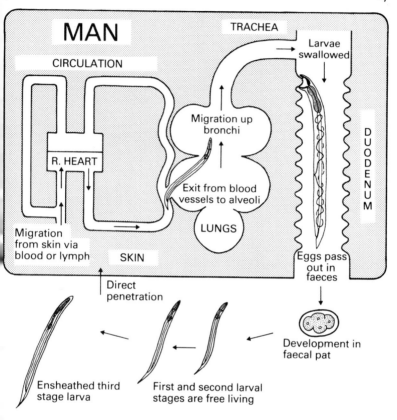

Fig. 2–8   Life cycle of *Ancylostoma duodenale*.

bronchi and trachea and swallowed. Upon arrival at the small intestine it
moults again to become the adult.

The mature hookworms, attached to the villi, feed by sucking host
blood. The amount of damage they cause depends very much on the
numbers present. Although they are small, an individual person can
harbour a thousand or more, and the estimated blood loss is about 200
ml/day. This leads to severe anaemia which in turn gives rise to numerous
secondary complications.

There is a whole range of genera of trichostrongyles parasitic in
domestic livestock, including *Haemonchus*, *Trichostrongylus*, *Ostertagia* and
Nematodirus. Their life cycles are identical with that of *Ancylostoma* except
for the mechanism of infection. The ensheathed third larval stage
migrates onto herbage and is accidentally swallowed by the grazing host.
The ingested larva is stimulated to exsheath in the stomach or intestine
and moults twice to become an adult. Some of these gut nematodes suck

blood; others damage the intestinal tissues by burrowing in the mucosa. The related nematode *Dictyocaulus* inhabits the lungs of livestock, causing parasitic bronchitis.

2.4.2 Ascarida

Ascarids are common intestinal parasites of vertebrates and *Ascaris lumbricoides*, the large roundworm of man and pig occurs world-wide, with an estimated 1000 million human cases. The mature worms are found in the small intestine and the females may measure up to 30 cm × 0.5 cm. A burden of only a few worms may result in literally millions of eggs being voided daily in the faeces. Larval development to the infective stage occurs within the egg, which may remain viable in the soil for months. Infection is thought to result most commonly from eating salad vegetables contaminated with eggs. The larva is stimulated to hatch by the conditions in the small intestine. Then, for reasons not understood, it penetrates the wall of the intestine and enters the hepatic portal vessels. It develops to the third stage in the liver and makes its way back to the gut via the heart, lungs and trachea. Development to maturity then occurs in the small intestine. The tissue stages of *Ascaris* may cause some damage, particularly if migrating in large numbers. However, the major effect of the worm is on host nutrition, and blockage of the intestine may occur, particularly in children.

There has been an increasing awareness in recent years of the possible hazards to health posed by another ascarid nematode, *Toxocara canis*, a parasite of the dog. The adult worms inhabit its small intestine and the life cycle is similar to that of *Ascaris* with a tissue-migration phase of long duration. Prenatal infection of puppies can occur when the larvae cross the placenta. Unfortunately the eggs of *Toxocara* are also infective to man and the larval worms undergo extensive migrations through the body. In particular, the eyes may be damaged, causing impairment or loss of vision.

The common pinworm of man, *Enterobius vermicularis*, is another ubiquitous ascarid. It parasitizes the large intestine and the female migrates at night to the anus to deposit eggs. The life cycle is direct, with infection resulting from ingestion of these resistant eggs. Unlike *Ascaris*, the larvae do not migrate through the tissues.

2.4.3 Spirurida

Spirurid nematodes are responsible for some particularly unpleasant diseases of man, including elephantiasis and river blindness. Their life cycles involve an intermediate host, frequently an insect. That of *Wuchereria bancrofti*, which together with the related *Brugia malayi* causes elephantiasis, is illustrated in Fig. 2–9. The mature worms, up to 10 cm long, are found in the lymph nodes and lymphatic vessels, particularly in the lower half of the body. In a proportion of patients the lymphatics

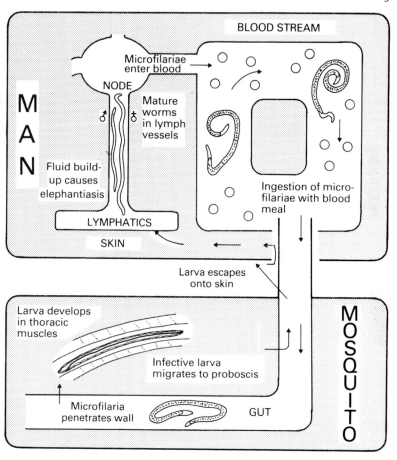

Fig. 2-9 Life cycle of *Wuchereria bancrofti*.

become blocked or atrophied and elephantiasis develops as a result of the build up of fluid in the tissues of the affected organ. The female worms are ovoviviparous, producing sheathed larvae called microfilariae. These pass through the lymphatics to the bloodstream and are taken up by various genera of mosquito vectors, with a blood meal. The larvae exsheath and migrate from the mid-gut to the thoracic muscles. Here they develop to the third larval stage and then migrate to the proboscis. The infective third-stage larva is thought to escape onto the host skin when the mosquito feeds, and then to enter via the bite wound. It migrates to the lymphatics and develops to maturity in about nine months.

River blindness is caused by the nematode *Oncocerca volvulus*. It is commonest in West Africa, the Volta basin having one of the most extensive foci, and has also been introduced into Central America. The mature thread-like worms, up to 50 cm long, are found in nodules of connective tissue beneath the skin. The females release unsheathed microfilariae which accumulate in large numbers in the vicinity of the nodule and also occur sparsely in peripheral blood. The intermediate hosts are blackflies of the genus *Simulium*. Development in the blackfly and the manner in which man is infected are similar to *Wuchereria* in the mosquito. *Simulium* is unusual among insect vectors of parasites in needing well-oxygenated running water for its larval development, which accounts for the distribution of *Oncocerca* along river systems.

In endemic areas, infection with *Oncocerca* begins in early childhood and there is a cumulative build-up in worm burden with increasing age. Blindness results when microfilariae invade the eye causing localized inflammatory lesions. In some areas more than 40% of the over-fifty age group may be blind and whole villages have been abandoned because of the disease.

The Guinea worm, *Dracunculus medinensis*, is a parasite of man in Africa, Western Asia and South America. The female worm, up to 1 m long, inhabits subcutaneous tissue. Its uterus becomes packed with larvae about 0.5 mm long. An irritating ulcer develops on the host skin in the vicinity of the worm's head. Contact with water stimulates the body of the worm to contract and expel a stream of larvae. If they are eaten by the crustacean *Cyclops* (cf. the tapeworm *Diphyllobothrium*) they will develop in its body cavity and man becomes infected by accidentally swallowing the *Cyclops* in drinking water. The larva makes its way to the subcutaneous tissues by an unknown route and develops to maturity in about one year. The chief harm of infection with *Dracunculus* results from secondary bacterial contamination of the skin ulcer. It is common practice to extract the worm from the wound over several days by carefully winding it out around a stick.

2.4.4 Trichinellida

Trichinella spiralis, parasitic in the intestine of many mammals, is the most important member of this group. The ovoviviparous female worms burrow into the mucosa and release larvae into the bloodstream. These larvae become encysted in muscles throughout the body and remain infective for long periods. Infections in man are generally thought to be acquired by eating contaminated pork. The migration of the larvae produces characteristic symptoms including fever and swelling of the face. The larvae cause tissue degeneration, and death can result from massive invasions.

Another trichinellid of man, *Trichiuris trichiura*, commonly inhabits the large intestine. The life cycle is direct with infection resulting from ingestion of resistant eggs. The chief symptoms associated with a large worm burden are anaemia and diarrhoea.

3 Physiological Adaptations for Transmission

3.1 Problems of transmission

Parasites are confronted with a special, but by no means unique, problem in reproducing the species. The potential habitat to be colonized is a discrete isolated unit, the host. For metazoan parasites the problem is compounded because multiplication does not occur in the vertebrate and can only be achieved by exit to the external environment. The term transmission describes this process of transfer from one isolated host to another. The evolution of parasitism is as much the evolution of efficient mechanisms for transmission as it is the adaptation to a life inside, or attached to, another animal.

In this chapter the physiological adaptations of the parasite associated with transmission are examined. For convenience, the infection of a vertebrate final host is regarded as the goal of the parasite. This is the host in which sexual reproduction normally occurs. However, there are many exceptions, particularly among protozoans, where sexual reproduction is absent or occurs in an invertebrate intermediate host. It is worth emphasizing that for the parasite, exit from or entry into a host is a violent transition between two very disparate environments. One of the major adaptations to parasitism has been the evolution of mechanisms to minimize the impact of this transition.

3.2 Resistant stages

A few parasites are transmitted from one host to another in an unspecialized and unprotected state, for example trophozoites of the protozoan *Trichomonas*, transmitted by direct contact. The commonest way of making the exit from or entry into a host is as a resistant stage. The parasite is protected during the transition from one environment to another, but this is in itself a problem as the structures which exclude hostile influences also confine the parasite. Its escape may be entirely dependent on the conditions provided by the host. Alternatively the parasite may initiate its own escape following activation by host-generated signals.

3.2.1 Formation, structure, and resistance

Cyst formation by intestinal protozoans is difficult to investigate due to their small size and the problems of isolating the cells from intestinal contents. The trophozoites of *Entamoeba histolytica* round off to secrete a

thin proteinaceous cyst wall. Within this, the parasite undergoes two
nuclear divisions and a glycogen vacuole and aggregations of RNA appear
in the cytoplasm. The cyst wall appears to protect the enclosed parasite
from chemical rather than physical influences.

The oocyst of coccidians such as *Eimeria* (Fig. 3–1) is one of the most
resistant structures known. It is formed from two types of inclusion
contained within the cytoplasm of the zygote and is a bilayer 0.25 μm
thick. A specialized plug, the micropyle, is present at one end. The mass of
cytoplasm within the cyst condenses and undergoes sporulation

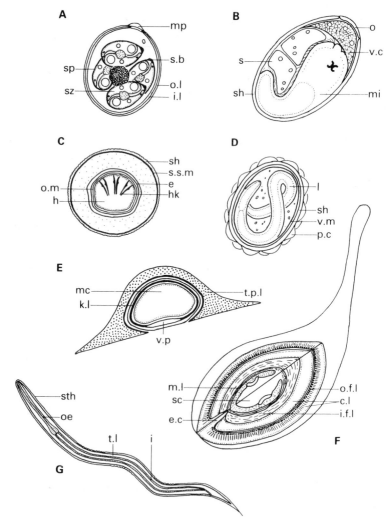

(sporozoite formation) which culminates in the formation of four sporocysts each with a special plug, the Steida body.

Conventional disinfectants have no effect on sporulated oocysts which can be stored in solutions of sulphuric acid or potassium dichromate sufficiently concentrated to kill all faecal microbes. The outer layer of the oocyst wall, composed of quinone-tanned protein, protects the contents against mechanical damage. The inner layer, composed of a protein matrix impregnated with lipid is the barrier to noxious chemicals. Only gases such as methyl bromide, ammonia and carbon disulphide are able to penetrate through the wall to attack the parasites within.

The eggs of parasitic flatworms are ectolecithal, the nutrients which sustain the developing embryo being contained in vitelline cells. Those of the digeneans such as *Fasciola* have a hard shell composed of tanned protein which originates principally as a secretion of the vitelline cells. In some species there is an operculum at one pole to permit escape of the larva. The digenean egg may develop to maturity in the uterus of the worm, in the tissues of the host, or in the external environment, depending on the species. The fully formed egg of *Fasciola* (Fig. 3–1) contains a mature miracidium larva, a viscous cushion and two or three fluid-filled sacs. Water and oxygen are required for development. The egg shell provides protection against mechanical damage and microbial attack. It is, however, completely permeable to water and the egg is unable to resist desiccation.

The structure, development and resistant properties of 'primitive' tapeworm eggs are similar to those of digeneans. The eggs of *Hymenolepis*, which develop *in utero*, are more complex. The hexacanth larva within the egg of *Hymenolepis diminuta* (Fig. 3–1) is surrounded by several concentric layers of material. From inside to out these are an oncospheral membrane, a proteinaceous embryophore, a cytoplasmic layer, a sub-shell 'membrane' and finally a tanned protein shell. The egg is able to

Fig. 3–1 Resistant stages (not to scale). (A) The sporulated oocyst of *Eimeria* (after LEVINE, N. D., (1961). *Protozoan Parasites of Domestic Animals and Man*. Burgess Publishing Co., Minneapolis.). (B) The egg of *Fasciola hepatica*. (C) The egg of *Hymenolepis diminuta*. (D) The egg of *Ascaris lumbricoides*. (E) The metacercarial cyst of *Fasciola hepatica* (after DIXON, K. E. (1965). *Parasitology*, 55, 215–26). (F) The cysticercoid larva of *Hymenolepis diminuta*. (G) The third stage ensheathed larva of the cat hookworm *Ancylostoma tubaeforme*. c.l – cytoplasmic layer, e – embryophore, e.c – escape channel, h – hexacanth, hk – hooks, i.f.l – inner fibrous layer, i.l – inner layer, i – intestine, k.l – keratin layer, l – larva, mc – metacercaria, mi – miracidium, m.l – membranous layer, mp – micropyle, o – operculum, o.f.l – outer fibrous layer, o.l – outer layer, o.m – oncospheral membrane, oe – oesophagus, p.c – protein coat, s – sac, s.b – Steida body, sc – scolex, sh – shell, s.s.m – sub-shell membrane, sth – sheath, sp – sporocyst, sz – sporozoite, t.l – third larval stage, t.p.l – tanned protein layer, v.c – viscous cushion, v.m – 'vitelline membrane', v.p – ventral plug.

withstand extremes of pH and to resist desiccation down to 20% relative humidity. The sub-shell membrane, in reality a dense layer 1μm thick, is believed to be the permeability barrier.

Many parasitic nematodes also make their exit from the vertebrate host as fertilized eggs (Fig. 3–1). These eggs have a simple and relatively uniform structure in the different groups. The fertilized ovum secretes three successive layers of material, the outermost of which probably corresponds to the fertilization membrane. The central layer is a hard shell composed of protein and chitin. The innermost layer is of predominantly lipid composition, up to 2 μm thick, and functions as the main permeability barrier of the egg. The eggs of some nematodes have an additional peripheral layer secreted by the uterus. Development of many nematode eggs to the first larval stage takes place in the external environment and those of species such as *Ascaris* have great powers to resist desiccation and other hostile influences.

Some parasites which exit from the vertebrate final host within an egg or cyst gain entry to a new host in that same state. Others, however, undergo a process of development either in the external environment or within an intermediate host, culminating in the formation of a resistant infective stage. The examples considered here are the digenean metacercarial cyst, the tapeworm cysticercoid larva, and the ensheathed third-stage larva of strongyle nematodes.

Most digeneans, other than schistosomes, are ingested into the vertebrate final host within a metacercarial cyst, formed following penetration into the tissues of a second intermediate host. In species such as *Fasciola*, where the cyst is formed on vegetation, it is exposed to the rigours of the external environment. The cercaria of *Fasciola* possesses four groups of gland cells (the cystogenous glands) in different parts of its body. In the process of cyst formation the cercaria attaches to the substratum by its ventral sucker, sheds its tail and secretes the contents of the cystogenous glands in sequence, to produce a multi-layered cyst. The two outermost layers (Fig. 3–1), thought to protect the parasite from mechanical damage and microbial attack, are incomplete and consist of tanned protein and mucoprotein respectively. The third layer has a complex structure and is composed of mucoproteins and mucopolysaccharides. The innermost layer, the permeability barrier, is composed of sheets of keratin-like material. The infective metacercaria is able to survive at low relative humidities for some time but the extent to which it can resist sub-zero temperatures is unknown.

The cysticercoid larva of *Hymenolepis diminuta* (Fig. 3–1) develops and becomes infective in the body cavity of its beetle host. In consequence, the hazards which it faces before reaching the site of parasitization in the duodenum of the rat host, are encountered in the alimentary tract. The fully formed larva is composed entirely of living tissues organized to protect the delicate scolex which lies at the centre. The scolex is surrounded by a membranous capsule consisting of thirty or more closely applied unit membranes bounded by a layer of collagen-like fibres. External to this is a layer of soft tissues bounded by another fibrous matrix. The outermost part of the cysticercoid is covered by a further layer of soft cytoplasmic tissue. The complete structure is able to withstand compressive forces up to 200 g before rupturing, presumably a protection against mastication when the beetle host is ingested by the rat. The inner membranous capsule protects the scolex from the low pH and proteolytic enzymes of the rat stomach.

When the second stage larva of strongyle and trichostrongyle nematodes moults to become the infective third stage, its cuticle separates from the underlying hypodermis forming a protective sheath, but is not cast off (Fig. 3–1). There is experimental evidence that this 'extra cuticle' protects the larva against desiccation at relative humidities greater than 50%.

There are exceptions to any generalizations that can be made about resistant stages. Nevertheless, there are features common to most eggs, cysts, etc., mentioned above. These include layers to protect against mechanical damage and microbial attack. The major physical threats to the enclosed parasite are extremes of temperature and desiccation the latter countered by permeability barriers of keratin-like or lipid material.

3.2.2   Quiescence

Development of the parasite within its protective capsule occurs to the point at which it is either infective to a potential host, or ready to escape into the external environment. The culmination of development is a period of quiescence characterized by a marked decline in activity and metabolic rate which persists until such time as the parasite is activated. Quiescence is of obvious value to an infective parasite with limited energy reserves, prolonging its viability and infectivity.

The biochemical basis of quiescence remains an enigma in spite of attempts to identify the precise point in the pathways of energy metabolism at which control is exerted. A detailed study of the phenomenon has been made in the egg of *Ascaris*. An examination of the activity of rate-limiting enzymes showed that, when the larva becomes quiescent within the egg, there is no rapid drop in the catalytic capacity of the pathways of glycolysis, the tricarboxylic acid cycle, or β oxidation.

The decline in metabolism might also be the result of changes in the specific activities of enzymes, substrate availability, or the inhibition of regulatory enzymes. Both quiescence and activation in *Ascaris* are accompanied by changes in the steady-state levels of intermediary metabolites. In particular, the ratio of ATP to ADP increases with the onset of quiescence and falls again on activation. Such changes could be the prime cause or merely the result of quiescence. Another factor which has been implicated is the compartmentation of substrates and enzymes either between or within cells. Unfortunately the available techniques for studying both enzyme activity and metabolite levels do not readily permit its existence to be demonstrated.

3.2.3   Activation and escape

The majority of encapsulated parasites require an external stimulus to promote activation and escape. Those considered here escape either into the external environment or into the lumen of the gut. The mechanisms of activation and escape are tailored to the particular site.

In those situations where the encapsulated larvae escape to the external environment, hatching is to a degree spontaneous. The miracidium larva of *Fasciola* is activated by light and possesses eyespots which presumably transduce the stimulus via its nervous system. A similar stimulus causing the hatching of strongyle nematode eggs has not been identified.

The hatching mechanisms employed by these two examples apparently depend on the difference in osmotic pressure between the egg contents and the external freshwater environment. In *Fasciola* the viscous cushion, lying beneath the operculum, hydrates and swells. The operculum is blown off, the cushion streams out and the miracidium is expelled by the expansion of the fluid-filled sacs. The entry of water into the egg of a strongyle nematode, which lacks an operculum, causes a rise in the internal pressure, rupturing of the shell, and escape of the larva. Both the miracidium and the strongyle larva apparently initiate the permeability changes, possibly by means of enzymic secretions.

The eggs of the digenean *Dicrocoelium* are swallowed by its snail host. The miracidium larva is activated by physico-chemical factors in the snail gut, particularly pH and reducing conditions. Both a temperature of 38°C and aerobic conditions are inhibitory, possibly devices to prevent premature hatching either before the egg leaves the gut of the vertebrate host, or in the external environment. The actual hatching mechanism may involve the release of enzymes by the miracidium to digest the opercular seal of the egg from within. The stimulus which activates *Dicrocoelium* is not specific, and hatching will occur in the guts of non-host snails and possibly other invertebrates.

The eggs of the tapeworm *Hymenolepis diminuta* are swallowed by its beetle host. The sub-shell membrane is ruptured by the masticating action of the beetle mouthparts causing immediate activation of the hexacanth larva. This activation presumably results from diffusion of a stimulant into the egg or of an inhibitor out of it. The larva is unable to escape by its own efforts and is released by the action of proteolytic enzymes present in the insect mid-gut.

Those parasites which escape from capsules into the lumen of the vertebrate gut may similarly depend on the external conditions. Two constituents of the vertebrate gut, bile salts and carbon dioxide, are of paramount importance and one or both have been implicated in the activation and escape of the remaining parasites considered here. Host-induced activation has obvious value for a parasite escaping into the intestine. During the course of evolution the parasite has become dependent on the host for stimuli which would normally be an endogenous part of its development. This dependence minimizes the chances of escape into an inappropriate situation. Resistant stages in the parasite life cycle have evolved both to protect the organism from contact with harmful external influences and to minimize the loss of water. Respiratory gases are apparent exceptions, able to pass freely through capsule walls. Carbon dioxide is present in the gut at high partial pressures and makes an ideal activator. Similarly bile salts, which function in the emulsification of lipids, are obvious agents to attack lipid-based permeability barriers.

When the larva of *Ascaris*, contained within the egg, and the ensheathed third-stage larvae of trichostrongyles are ingested into the mammalian gut, they are activated by carbon dioxide and a temperature of 37°C. The larva of *Ascaris* is stimulated to secrete enzymes which digest both the vitelline membrane and the egg shell, permitting it to escape. Tricho-strongyle larvae complete the process of moulting and escape from the sheath. The mechanism of exsheathment is controversial but may involve the release of hydrolytic enzymes to act at a ring of weakness on the inner surface of the sheath.

The metacercaria of *Fasciola* also requires carbon dioxide for activation in the small intestine, with a temperature of 37°C and reducing conditions. However, the larva can be held in the activated state for at least 24 hours without escaping from the cyst. Addition of bile salts alters the permeability of the cyst wall and the parasite then effects its own escape by digesting the ventral plug.

Escape of the sporozoites from a coccidian oocyst is a two-stage process. In the intestine of ruminants activation is stimulated by carbon dioxide which causes a change in the permeability of the micropyle cap, probably the result of the action of enzymes originating within the cyst. In poultry, oocysts may be activated either by carbon dioxide or by the mechanical action of the gizzard rupturing the cyst wall. Release of sporozoites from the enclosed sporocysts requires a temperature of 37°C, bile to alter the permeability, and trypsin to digest the sporocyst plug. The motile sporozoites then rapidly escape through the aperture.

The encapsulated stages of tapeworms depend entirely on the vertebrate host for release. Carbon dioxide does not appear to be necessary but the larvae require treatment with proteolytic enzymes such as pepsin and trypsin at a temperature of 37°C. The tissues of the cysticercoid of *Hymenolepis* are digested away to release the membranous capsule containing the inactive scolex. Bile salts alter the permeability of the membranes, bringing about the activation of the scolex, an event which may involve the diffusion of an inhibitory substance from the space between it and the capsule wall.

As with the phenomenon of quiescence, the related process of activation is poorly understood. Even where the stimulus (such as carbon dioxide) has been identified, its mode of action is uncertain. The phenomenon has been most extensively investigated in third-stage trichostrongyle nematodes where the action of carbon dioxide may be exerted either via a peripheral receptor (a sensory receptor?) or directly on the metabolism of the larva. It has been suggested that one of the processes triggered by the stimulus is the secretion of moulting fluid from the 'excretory glands'. This may be controlled by the release of neurosecretory hormones in the vicinity of the glands. The exact nature of the various compounds involved remains to be discovered.

3.3 Host location

The monogenetic and digenetic flatworms and the strongyle and trichostrongyle nematodes have free-living larval stages which actively locate prospective hosts. There have been innumerable descriptions of the behaviour of these larval parasites and particularly of the changes induced by stimuli of host origin. However, very few satisfactory explanations have been put forward of the significance for host location of these behavioural changes. A major cause of confusion has been the failure to distinguish between the different forms of orientation behaviour such as taxes and kineses, both of which can bring about the aggregation of animals but by very different mechanisms.

3.3.1 Miracidia and cercariae

Digenean miracidia are elongate organisms 100 to 200 μm long and covered with ciliated epithelial cells. The beating of these cilia propels the miracidium through the water at speeds up to 1.5 mm/sec but in those species so far studied in any detail there appears to be little or no control over the rate of ciliary activity. Swimming speed is relatively constant in the short term, turning is achieved by differential contraction of the musculature to bend the body and, unlike protozoans, reversal of the cilia is not possible. The majority of miracidia are relatively specific in their choice of gastropod mollusc host.

Digenean cercariae have a more diverse morphology and a greater range of behavioural patterns. This may be related to the fact that they infect a wider range of hosts including snails, insects, fish and mammals. Most cercariae possess a muscular tail, the lashings of which propel them through the water. The direction of locomotion is tail first, but in some species reversals are possible. Both direction and speed of swimming are more variable. Cercariae exhibit a variety of basic swimming movements. For example, the unstimulated cercaria of *Schistosoma mansoni* has a rhythmic pattern of upward swimming alternating with downward passive sinking which keeps it just beneath the water surface. At the other extreme cercariae may swim continuously in an apparently random manner.

Miracidia of different species respond to light (both direction and intensity) and to chemicals emanating from the snail host. The effect of chemical stimulation is almost invariably an increase in frequency of turning and of exploratory attachments to the substratum. It is also likely that mechanical stimuli play an important role in the penetration process.

Individual species of cercariae have been shown to respond to some or all of the following stimuli: light (direction and intensity), chemicals, vibrations, and temperature gradients. In schistosome cercariae, the period of active swimming may increase at the expense of the passive sinking phase. When the cercaria hits the water surface a sinking phase is not initiated and it continues swimming horizontally. Chemicals diffusing from host skin will initiate virtually continuous movement in a random direction and with frequent reversals. Cercariae show varying degrees of adaptation to the stimuli, some responses decaying in seconds, others persisting for minutes.

The ability to respond to extrinsic stimuli implies the possession of sensory receptors suitable for their detection. Investigation of larval structure by electron microscopy has revealed a wide range of peripheral nerve endings which, on grounds of comparative morphology, are presumed to have a sensory function (Fig. 3–2). The miracidium of *Fasciola hepatica* has at least six different types, situated around its anterior. In cercariae, the receptors are spread widely over the body and tail but with a concentration at the anterior. In *Schistosoma mansoni* there are at least 126 classified into four different types. These sensory structures typically consist of an axon terminating in a bulb of cytoplasm bearing one or more modified, and presumably non-motile, cilia. The cilia may project for several microns into the external environment. It has been suggested that the simple uniciliate endings function as mechano-receptors. Multiciliate nerve endings frequently lie at the bottom of sunken pits and have been tentatively identified as chemoreceptors. Other types may function as photoreceptors, proprioreceptors and in the detection of gravity.

Apparently then, miracidia and cercariae have the capacity both to perceive, and respond to their environment. They exhibit complex behavioural patterns which can be modified by host-generated stimuli. As yet, there is no satisfactory explanation of the way these changes in behaviour improve the parasite's chances of contact with the host. Analogies have been drawn between parasite-host and predator-prey interactions in which the patterns of behaviour exhibited by parasite larvae are interpreted as ways of conserving limited energy reserves whilst maximizing the probability of host contact. There are two alternative

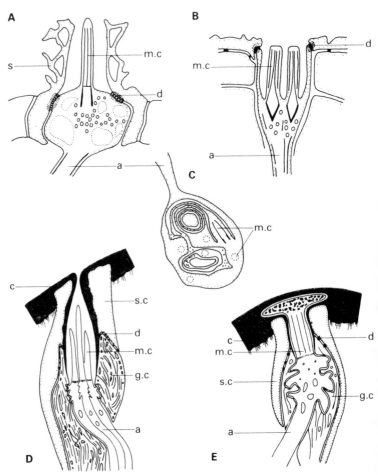

Fig. 3–2 Peripheral sense organs. **A–C** from the miracidium of *Fasciola*. (**A**) Sheathed uniciliate ending. (**B**) Multiciliate pit receptor. (**C**) Internal club-shaped ending. (**D**) Amphid of the filarial nematode, *Dipetalonema*. (**E**) Anterior papilla of *Dipetalonema*. (**D** and **E** after McLAREN, D.J., in CROLL 1976.) a – axon, c – cuticle, d – desmosome. g.c – gland cell, m.c – modified cilium, s.c – support cell, s – sheath.

strategies: 'sit and wait' and 'active searching'. A fast-moving miracidium, locating a slow-moving snail host, is presumably using an active searching strategy. Continuous swimming in a random direction may take the miracidium into the vicinity of a potential host. Stimuli generated by the host will then cause increased turning which serves to keep the miracidium in proximity to the snail, so enhancing the chances of contact. In this interpretation of miracidial behaviour it is unnecessary to postulate a taxis to the snail. The pattern of alternate sinking and swimming employed by the cercaria of *Schistosoma mansoni* maintains it in a favourable location for host contact. This presumably exemplifies a 'sit and wait' strategy in which energy expenditure is minimized when the host is not in the vicinity. When the human host comes wading through the water, turbulence, shadows and skin chemicals will all increase swimming activity which subsides back to basal levels if the host moves away before contact has been made. Again, it is not necessary to postulate a taxis to host skin as the increased but random activity should enhance the chances of contact. The obvious benefits of the strategy are to produce maximum effort only for the brief period when the host is in the vicinity of the larva.

3.3.2   Larval nematodes

Both the trichostrongyle nematodes of grazing animals and the hookworms spend their first two larval stages feeding on bacteria in the faecal pats in which they hatched. To infect the host they must migrate out of the faeces. The trichostrongyles are thought to migrate onto herbage and infection results from their accidental ingestion. Hookworms migrate less onto herbage and are said to accumulate on prominent objects just above the soil surface. They exhibit nictitating behaviour, standing on their tails in an erect posture waving the anterior of the body from side to side. Infection of the host is by direct penetration through the skin surface.

Dispersal of infective nematode larvae is obviously important, but the actual distances moved appear to be very small. (A miracidium may travel 50 to 100 times faster than a larval nematode.) Some novel mechanisms of dispersal have been described. The larvae of *Dictyocaulus viviparus*, *Cooperia* and *Trichostrongylus* parasitic in cattle, climb up the sporangiophores of the fungus *Pilobolus* which grows on the faecal pat. They may then be fired up to 3 m when the sporangium explodes. Larvae of other nematodes (*Ostertagia*, *Oesophagostomum*) may be dispersed attached to the legs of dung flies.

Nematodes require a film of moisture for locomotion, the basic pattern of movement being an undulating track with a marked unilateral bias. Forward movement is punctuated by reversals and changes in direction such that the larvae move in a series of arcs. In the first and second larval stages, feeding on bacteria, the activity is spontaneously generated but can be modified by external stimuli. Reversals are frequent and dispersal of the larvae is thereby limited. The activity of the infective third stage is much reduced (an energy-conserving mechanism) but the larva remains responsive to external stimuli. It undergoes a burst of locomotory activity of sufficient duration to permit exploitation of the stimulus.

Different species of larval nematodes have been shown to respond to a wide range of stimuli, including light, temperature, chemicals and gravity. As with digenean larvae there is considerable confusion as to whether the responses are taxes or kineses.

Larval nematodes possess a variety of presumed sensory endings of which two types, amphids and anterior papillae (Fig. 3–2) appear to be ubiquitous. The amphids are located on either side of the mouth and each consists of an invaginated channel lined with cuticle and opening to the exterior at a small pore. There are three constituent cell types, a supporting cell around the cuticular channel, a secretory cell and several sensory receptor cells. Each receptor consists of an axon terminating at one or more modified cilia. Experimental work employing behavioural mutants which do not possess the ability to detect chemicals, suggests that amphids are chemoreceptors. Anterior papillae occur around the nematode mouth and have the same basic arrangment of three cell types. The gland is less well-developed and the modified cilia may terminate beneath the cuticle (Fig. 3–2). These papillae are generally considered to be mechanoreceptors.

Although elements of the behaviour of strongyle and trichostrongyle nematodes have been described, no satisfactory explanation has been advanced which correlates the observed changes with enhanced host contact. It has been suggested that responses to the macroenvironment take the larva into the 'orbit' of its host. An example of this would be the kinetic responses of trichostrongyles to light, causing migration up grass blades. Once the larva has reached the host orbit it can respond to host-generated stimuli, for example the movement of hookworm larvae up temperature gradients such as might be produced by host skin.

3.4 Penetration mechanisms, metamorphosis and growth

The process of transmission does not necessarily end with location of, ingestion by, or inoculation into a suitable host. The parasite may have to penetrate into or through tissue barriers to reach the site where development occurs.

The sporozoites of coccidia escape from their cysts into the lumen of the gut and penetrate intestinal epithelial cells. Those of *Plasmodium* are injected into the bloodstream and rapidly penetrate cells such as hepatocytes. Merozoites, the products of asexual multiplication in both coccidia and *Plasmodium*, also reinvade tissues to develop further. It has been suggested that there are receptors on the anterior end of sporozoites and merozoites which interact with molecules on the surface of the host cell, ensuring a degree of specificity in the type of cell penetrated.

The process of penetration takes only a few seconds. The conoid apparatus of the parasite is probably involved, together with secretions from the rhoptries (Fig. 2–1). Unfortunately, the minute size of these parasites makes experimentation difficult and explanations of penetration are speculative. It has been variously suggested that penetration is mechanical, that it involves hydrolytic enzymes, and that the configuration of the host cell membrane may be modified by surface active agents, to admit the parasite. It is thought that the merozoite of *Plasmodium* stimulates the host erythrocyte to phagocytoze it. Phagocytosis is not a normal activity of such cells.

Direct and active entry into the tissues of the host via the body surface is not a particularly common route presumably because, in mammals at least, one of the functions of the surface is to keep out pathogens. On the basis of morphological evidence rather than biochemical studies, it has been concluded that digenean miracidia gain entry into the tissues of the snail by means of hydrolytic secretions. The process of penetration takes from 5 to 30 minutes depending on the species. The miracidium of *Fasciola* possesses a large flask-shaped apical gland and two pairs of smaller accessory glands, all of which open at the extreme anterior of the apical papilla (Fig. 3–3). The secretions of these glands probably help to stabilize attachment as well as lysing the snail epithelium.

The cercaria of *Schistosoma* and the third larval stage of hookworms have to penetrate the more formidable barrier of mammalian skin. The schistosome cercaria is equipped with two pairs of gland cells situated

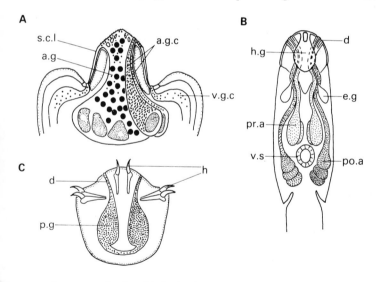

Fig. 3–3   Secretory glands of infective larvae. (A) Miracidium of *Fasciola*. (B) Cercaria of *Schistosoma mansoni* (after STIREWALT, M. A. (1974). *Advances in Parasitology*, 12, 115–82). (C) Hexacanth of *Hymenolepis diminuta*. a.g – apical gland, a.g.c – accessory gland cells, d – duct, e.g – escape gland, h – hooks, h.g – head gland, p.g – penetration gland, po.a – postacetubular gland, pr.a – preacetubular gland, s.c.l – surface cytoplasmic layer, v.g.c – vesiculated gland cell, v.s – ventral sucker.

anterior to the ventral sucker and three pairs posterior to it. The content of these glands are secreted via ducts which open at the anterior of the larva (Fig. 3–3). The posterior set of glands secrete principally mucus to assist adhesion of the larva's suckers to the skin. The secretion of the anterior set contains calcium and proteolytic enzymes, the latter capable of digesting the intercellular 'cement' of the horny layer and of the living epidermis. The process of penetration to the basal lamella of the Malpighian layer takes only a few minutes.

Some species of hookworm larvae possess gland cells in the pharynx the secretions from which pass out through the mouth. These are thought to contain hydrolases which digest skin components, permitting entry of the larva. However, other species may pass through the skin using only simple mechanical activity and their pharyngeal glands are much smaller.

The more common route of active entry into the body tissues is via the gut wall, presumably facilitated by the secretion of hydrolytic enzymes. The hexacanth larva of tapeworms combines secretion with mechanical activity to penetrate gut tissues. The larva possesses three pairs of hooks, associated musculature and penetration glands (Fig. 3–3). On activation in the lumen of the gut, the hooks commence a rhythmic sequence of extension and retraction with a periodicity of a few seconds. The activity of the larva is not directional, but the tearing action of the hooks supplemented by the lytic action of the secretions result in migration through the tissues. The whole process appears to be the automatic consequence of activation.

The miracidium and cercaria larvae of the digenetic flatworms are adapted for an active swimming and non-feeding existence. Penetration of the host is accompanied by a major reorganization of parasite tissue and physiology which can properly be described as a metamorphosis. The miracidium larva sheds its ciliated epithelial cells and secretes a syncytial epidermis more suited to the absorption of nutrients. The cercaria of Schistosoma undergoes a similar metamorphosis, characterized by change in the nature of the surface membranes, loss of gland cells, increase in osmotic sensitivity to water and activation of its intestine. Metamorphosis of both larvae can be brought about in vitro, chemical stimuli of host origin being prerequisites. A variety of skin lipids are the most likely stimulants for schistosome cercariae, but the nature of those acting on miracidia is still controversial.

The transmission stage, be it quiescent within a cyst or a free-living larva, is not capable of growth until it has gained access to a host. It seems likely that the physical and chemical conditions provided by that host act as the trigger for growth. One probable determinant of host specificity is the subtle variation in the environmental conditions provided by closely related hosts. In at least one parasite, Schistosoma, the initiation of growth is separated from the process of metamorphosis. The latter is complete within one or two days after penetration, but growth does not commence until the parasite arrives at its site of parasitization in the hepatic portal system up to twenty days later. The nature of the stimulus which initiates growth is unknown. It seems likely that other examples of the temporal separation of metamorphosis and growth will be found in metazoan parasites which undergo prolonged tissue migration to reach the site of parasitization.

3.5   Host-parasite synchronization

A number of phenomena have been described in which the life cycle of a parasite is synchronized with some aspect of the life of the host, to facilitate transmission. Two examples, the periodicity of microfilariae in peripheral blood, and the arrested development of nematodes after infection, are discussed below.

The microfilariae of spirurid nematodes such as *Wuchereria* are numerous in the peripheral blood for only part of the day. This periodicity is related to the feeding activity of the insect vector and the actual pattern depends on the species. Microfilariae are carried passively round the body in the bloodstream, completing the circuit about once every minute. As numbers of larvae diminish in the peripheral blood there is a corresponding build up at the arteriolar end of the pulmonary capillaries. This rise occurs because the microfilariae are stimulated to swim against the flow, instead of moving passively with the blood through the lungs. It has been suggested that their chemoreceptors may directly monitor oxygen tension in order to trigger the swimming activity. Oxygen tension in the lungs fluctuates in relation to the pattern of activity and rest of the human host. It is by no means obvious why it is advantageous for a microfilaria to remain in the pulmonary capillaries at times of the day when the insect vector is not feeding.

The arrested development of larval nematodes in the final host is yet another example of a quiescent phase in the nematode life cycle (cf. the larva of *Ascaris* within the egg, or the ensheathed trichostrongyle larva). It might be supposed that the best strategy for a parasite which has infected a host is to develop rapidly to maturity and start to produce eggs. The arrested development of trichostrongyle nematodes postpones maturation and egg laying of parasites acquired in the autumn to the following spring. The adult life of trichostongyles may be as short as 2–3 weeks and the larvae on the pasture are very susceptible to extremes of temperature and desiccation. Maturation and egg laying in the autumn would therefore largely be wasted.

The causative factors are of two kinds: environmental influences such as temperature which act on the larvae *before* infection, and the suitability of the host as a habitat for the parasite. The mechanism which arrests and restarts development is not understood. The stimuli for resumption may be endogenous or of host origin.

4 Epidemiology

4.1 What is epidemiology?

The aim of epidemiology is to describe and explain the parasite life cycle in quantitative terms. This means more than simply making a record of the number of infected hosts, the number of parasites they harbour, etc. Epidemiology is concerned with the variation in levels of infection, with time and from place to place. The quantitative approach necessarily involves a mathematical formulation of the problem under investigation. A grasp of calculus is all the mathematics required for most purposes. Nevertheless parasitologists have been resistant both to the ideas and the methodology which the approach entails. The concepts have been applied in their most highly developed form to the study of malaria, and some progress has been made in describing the dynamics of schistosomiasis, fascioliasis, and diseases caused by trichostrongyle nematodes. Other parasites have received only cursory attention and there has been a limited search for underlying theoretical principles.

4.2 The life cycle as a dynamic system

There are four factors which govern the size of any animal population: birth rate, death rate, emigration and immigration. More than fifty years ago the idea was conceived that the dynamics of a parasite life cycle could be described in terms of these four influences. The simplest qualitative analogue or model of the cycle is a flow diagram such as that illustrated in Fig. 4–1. The diagram depicts the one-host life cycle of the nematode *Trichuris trichiura*. There is no asexual multiplicative stage and no multiplication within the host. With slight elaboration the diagram could equally well represent the life cycle of a monogenetic flatworm, or other nematodes such as *Enterobius*, *Ascaris*, the hookworms and the trichostrongyles. For convenience the host population is treated as being stable, i.e. births equal deaths and there is no immigration or emigration.

The parasites are found in two situations, in the host population (as juveniles and adults), and in the external environment (as eggs). The rate of egg production by the adult parasites is the sole birth process. The exit or emigration of these eggs from infected hosts is simultaneously an immigration into the pool of parasites in the external environment. Similarly, the process by which further members of the host population are infected is a simultaneous emigration from the environmental pool and an immigration into the population of parasites established in the

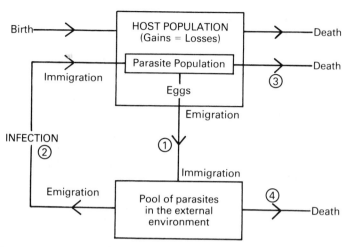

Fig. 4–1 Diagrammatic representation of the life cycle of *Trichuris trichiura*. The cycle can be represented as two emigration–immigration processes ① and ②, and two death processes, ③ and ④. (After ANDERSON, R. M. in KENNEDY, 1976).

hosts. Mortality of parasites occurs both in the host and in the external environment. The dynamics of this particular host-parasite interaction can thus be treated as a series of four rate-parameters.

Provided that realistic estimates of the four rates can be made, then the qualitiative analogue can be converted into a quantitative model. Such models are commonly formulated in one of two ways. In a deterministic model the life cycle can be depicted as a series of differential equations whose solution specifies the size of the parasite population. The values which individual parameters in the equations take, is fixed prior to computation. In a stochastic model the likelihood of some event such as a birth or death occurring, is described by a probability distribution. By selecting the value for a particular parameter from such a distribution, an element of variability (and perhaps realism) is introduced into the model.

The basic idea can be elaborated into a life cycle model of a parasite with two or three hosts in which the host populations are themselves capable of fluctuation. Such formulations are obviously considerably more complex.

4.3   Analysis of rates

It is not difficult to visualize a parasite life cycle in terms of its primary population processes. In practice, the problem is to obtain accurate estimates for individual parameters such as the rate of infection. To gather such information requires an integrated approach involving laboratory work, field work and mathematical analysis.

A parameter, such as the rate of infection, is itself obviously the product of an interaction between several factors. We must therefore break down the infection process into its constituent parts and study these individually. These elementary processes can be investigated under defined conditions in the laboratory, and where possible in the field. The information gleaned permits the experimenter to assign a numerical value to each constituent. A mathematical expression can then be derived, interrelating the various parameters to provide an estimate of the rate of infection. Total realism is obviously not feasible, but the expression will generally be considered adequate if it takes account of the major sources of variation. A vital part of the procedure is to check the predictions of the equation against observed rates of infection in the field. A single example will serve to illustrate this process.

One component of the inoculation rate in malaria is s, the proportion of mosquitoes with infective sporozoites in their salivary glands. (Inoculation rate does not mean the same as infection rate since a proportion of the people bitten will already harbour the malaria parasite.) This in turn is dependent on:

  a, the average number of blood meals on man taken by a mosquito per day;
  x, the proportion of bites on man which are infective to the mosquito (proportion of people with gametocytes in the blood);
  n, the time in days for malaria development in the mosquito;
  p, the probability of a mosquito surviving through one day.

Typical values for endemic malaria (in Madras, India c. 1940) were $a = 0.0125$ bites/day, $x = 0.065$, $n = 9$ days, and $p = 0.775$. These variables can be related in an equation (see MACDONALD, 1973, for its derivation) to give an estimate of s where:

$$s = \frac{p^n \cdot ax}{ax - \log_e p}$$

Substituting the numerical estimates of the various parameters gives $s = 0.032\%$. The rate for s in the field, obtained by the dissection of more than 10 000 mosquitoes, was 0.064%. To the sceptic the discrepancy between the predicted and observed results might seem large. To the epidemiologist their proximity suggests that the major causes of variation in mosquito infectivity are accounted for in the equation.

The method of combining variables in a simple algebraic expression which describes one of the primary population processes, is most useful where the total length of a parasite life cycle is short. In such a case, the fluctuations of environmental influences (e.g. temperature) are likely to be so small that they will have little effect. The over-winter cycle of *Fasciola hepatica* in the snail can take nine months, during which time the host and parasite are subjected to a wide range of temperatures. The models of fascioliasis which have been devised must therefore take account of these seasonal effects of temperature.

Another approach which has been attempted, by analogy with human demography, is the use of life tables. They are ideal for a population which is at equilibrium but cannot be readily adapted to cope with seasonal variations in birth and survival rates or with rapidly changing populations like those in epidemics. Life tables are therefore likely to be of limited use in the study of parasite populations.

4.4   Regulation of parasite populations

The pattern of a disease can vary in different geographical localities and at different times in the same locality. The term *endemic* is used to describe a situation in which, although there may be seasonal fluctuations, the pattern of disease is repeated each year. The level of endemicity, or prevalence, is not necessarily high in a stable disease situation. Conversely, the prevalence of some diseases like schistosomiasis can be greater than 90% in certain communities.

The term *epidemic* is used to describe a situation in which the level of disease shows haphazard and extreme variation. Malaria and fascioliasis are two examples which may show this characteristic. Epidemics are very dramatic but in the long term endemic disease may cause more damage, particularly if prevalence is high. It seems reasonable to suppose that endemic disease is regulated in some way to produce the stable pattern, but that epidemic disease is not subject to the same constraints.

We are accustomed to the idea that the operation of machines can be regulated by a process of feedback control. The governor of a steam engine or the thermostat of a central-heating boiler are just two examples. These controls operate by negative feedback, so that when the output of the machine exceeds the desired level the magnitude of the input signal is reduced, causing the machine to return to equilibrium. It is pertinent to ask whether negative feedback mechanisms operate in parasite life cycles to generate stability.

There is now considerable evidence of the way in which animal numbers are regulated, and a concise account can be found in the book on population dynamics by SOLOMON (1976). The major influences can all be classified as either density-independent or density-dependent. Environmental temperature and moisture are density-independent factors, acting on the primary population processes quite independently of population size. Density-dependent factors work in a more complex and subtle way, related to the number of individuals in the population. They provide a form of negative feedback, so that if numbers rise above or fall below equilibrium, then the population will be steered back to the equilibrium level. Predation and food supply are examples of such factors. The concensus of opinion is that density-independent factors cannot steer populations to equilibrium. If they were the only regulators then they would, sooner or later, cause extinction of the population.

Stability does not mean constancy; a delayed density-dependent process imposes oscillations round a mean level because of overshoot in the control process. The position of the equilibrium may itself also be continuously changing due to environmental influences.

A number of density-dependent processes have been described in parasite life cycles, and others will doubtless be identified as their importance in disease control is recognized. An example affecting the birth process is shown by *Fasciola*, parasitic in sheep, and probably many other intestinal helminths as well. The number of eggs produced per fluke decreases as the total fluke burden increases. Infection of the snail host of *Fasciola* (an immigration process) is also density-dependent. An increase in the density of miracidia results in a less than proportional increase in the parasitization of the snail host. Density-dependent mortality of nematodes in the intestinal tract, perhaps due to crowding at high parasite levels, has also been described.

The host immune response may also limit parasite populations in a density-dependent manner (see Chapter 7). There is in any locality a finite number of strains of malaria to which an individual is exposed and acquires immunity. The presence of a reasonable level of immunity in a human population produces a stable disease situation. Its absence, due to climatic variation, sporadic chemotherapy or vector control, can permit explosive epidemics such as have occurred in Sri Lanka and Guatemala in recent years. Concomitant immunity to schistosomes may also regulate the worm population. A moderate level of infection confers on the host a resistance to further invading cercariae.

The lethal level effect is another form of density-dependent regulation, probably unique to parasite populations. Where a parasite is sufficiently pathogenic, there will be a level of parasitization, the lethal level, which kills the host. Many parasite species are distributed through the host population in an overdispersed manner, best described by a negative binomial distribution (KENNEDY, 1975). A large proportion of the parasites is present in a small proportion of the hosts. This will tend to magnify the impact of the lethal level because the death of a heavily infected host removes a disproportionately large number of parasites from the population. A classic example of the lethal level in action is seen in acute fascioliasis of sheep, where death results from a worm burden in excess of 250 flukes per host.

4.5   Uses and limitations of models

Epidemiological models can be put to a variety of practical uses. For example, the sensitivity of an endemic disease level to changes in any or all of the controlling factors, can be assessed. Such exercises can provide answers to questions like 'by how much do we need to reduce the average life span of a mosquito in order to cause cessation of transmission?'.

As well as providing insights into the dynamics of transmission, models can be used in schemes for disease control and eradication. Malaria is the only disease caused by a parasite in which any real progress has been made towards this elusive goal. The models of malaria have been used to analyse field data on endemic and epidemic disease patterns. They allow alternative control strategies to be evaluated at negligible cost before large sums of money are committed in control programmes. For example, the models can indicate the frequency of drug administration to the human population required to achieve control in a particular period of time. A yardstick of expected progress in a control programme can be established, against which actual progress can be compared.

Knowledge of other diseases is not based on such sound theoretical principles as malaria and control measures are frequently empirical. For detailed information on their epidemiology the reader should consult the various monographs cited in the references.

Another area in which models have been used is in the prediction of likely disease incidence. In temperate regions climate can vary markedly from one year to the next. A knowledge of the climatic conditions most likely to generate an epidemic, for example of fascioliasis or parasitic gastroenteritis in livestock, permits a forecast of incidence to be made. The farmer can then take appropriate action such as spraying snail habitats with molluscicide or dosing livestock with anthelmintics.

An increasing awareness of the usefulness of models makes it likely that more sophisticated versions will be developed to assist our understanding of transmission and to facilitate the control and eradication of disease.

5 Physiology and Metabolism of Endoparasitic Stages

5.1 The host as an environment

Parasites are adapted to exploit the rather unusual environment provided by the interior of another living organism. Before some of these adaptations are described, the advantages and disadvantages likely to accrue from such a way of life are examined. Only the range of environments provided by a mammal host are dealt with, but a similar exercise could equally well be carried out for hosts as diverse as insects or molluscs.

The first and most obvious consideration is one of space. It would be convenient if we could make the generalization that large parasites are more likely to be lumen rather than tissue dwellers. Unfortunately some of the larger nematodes such as *Dracunculus* are found in connective tissue. Nevertheless, tissues are likely to provide a more restricted site than the lumen of an organ. The flow of fluids through organs creates particular problems for lumen dwellers. Position must be maintained against the flow by the use of suckers or hooks which attach the parasite to the wall of the organ (e.g. schistosomes or tapeworms) or by active swimming against the stream (e.g. intestinal nematodes).

The physiological processes occurring within a mammal maintain a stable internal environment with consequent benefits for an endoparasite. Temperature, osmotic pressure and ionic composition are regulated within narrow limits and body fluids are generally well buffered with consequent stable pH. The major difference in composition is between potassium-rich intracellular and sodium-rich extracellular fluids. The lumen of the intestine is an exception to this stability, composition being in part dependent on diet. The stomach has a low pH due to the secretion of hydrochloric acid, and the pH of the duodenum may fluctuate, particularly during the passage of food from the stomach.

In contrast to the stability of the ionic and osmotic environment, the tension of respiratory gases varies widely in different tissues. The oxygen tension (pO_2) of arterial blood is around 100 mm Hg, that of venous blood may be as low as 5 mm Hg. The oxygen supply to tissues depends on the degree of vascularization and the proximity of blood vessels. The partial pressure of carbon dioxide (pCO_2) is seldom lower than around 40 mm Hg and can be much higher in actively respiring tissues. The exact gas tension found in the lumina of organs is controversial. It was formerly thought that the pO_2 was very low in the intestine, apart from in those regions adjacent to the tissues. However, recent work suggests that it may be as high as 50 mm Hg, due to the stirring action of peristalsis. All workers seem agreed that the pCO_2 is high, frequently reaching tension greater than 100 mm Hg.

The major benefit which a mammal host confers on a parasite is a supply of nutrients. The relationship between some parasites and their hosts has been likened to that between an organ and the body as a whole. The levels of ions, metabolites and nutrients, in the tissues of the parasite may be in equilibrium with those of the host blood stream or intestinal lumen. Intestinal dwellers get a frequent supply of concentrated and digested food. The level obviously fluctuates between meals, but the contribution of intestinal secretions, and the leakage of small molecules such as amino acids from the tissues, ensures the presence of some nutrients even in a fasting host. The bloodstream is also a constant source of nutrients. The levels in peripheral blood (and hence in the interstitial spaces) are controlled by the homeostatic action of the liver. Those in hepatic portal blood can vary greatly as absorbed compounds are transported to the liver following a meal.

A strategy used by some parasites is to treat the host as either a renewable or non-renewable resource simply by consuming the tissues. Parasites which have adopted this means of exploitation include *Plasmodium*, *Schistosoma*, and the hookworms, all of which feed on blood cells.

5.2   Nutrition

Animals use a variety of mechanisms, either singly or in combination, for the uptake of nutrients. The simplest is passive diffusion into the tissues down a concentration gradient. The rate of uptake is dependent on the slope of the gradient, which means that the cell or organism must be bathed in a rich soup of nutrients for appreciable transport to occur. Mediated transport takes place against the concentration gradient and requires an energy source. The mechanism is thought to involve 'carrier molecules' in the plasmamembrane which become saturated at high nutrient concentration, resulting in a transport maximum. A further mechanism, common in lower animals, is uptake by endocytosis. This term covers both particles (phagocytosis) and fluids (pinocytosis). Endocytosis involves the invagination of a pocket of plasmamembrane into the cell to form a food vacuole. Primary lysosomes fuse with the vacuole to form a phagosome in which digestion takes place. The nutrients diffuse into the cytoplasm and the undigested residue is defaecated from the cell.

Free-living heterotrophic Protozoa employ endocytosis as the chief feeding mechanism and this has persisted in the endoparasites. *Entamoeba histolytica* feeds on tissue debris and erythrocytes, and the bloodstream forms of *Trypanosoma* feed on blood fluids. In the flagellar pocket, trypanosomes have either a specialized cytopharynx terminating in a cytosome, or an unspecialized region of the pellicle across which endocytosis occurs.

The merozoites of *Plasmodium* probably do not feed, but following invasion of cells they transform into trophozoites. The cytostome becomes functional and the contents of the erythrocyte are taken in by endocytosis. A characteristic pigment, the by-product of digestion, is deposited in the cytoplasm.

There is some evidence that the endoparasitic Protozoa can take up nutrients across the bounding plasmamembrane by diffusion or mediated transport. There is no clear pattern between species, but substances transported in this way include sugars and amino acids. It has been suggested that these parasites employ mediated uptake for specific nutrients which are required at higher levels in the cytoplasm than could be maintained by passive diffusion or endocytosis.

Digenean flatworms have a well-developed, bifurcated gut which is blind-ending and functions as a tidal system. They feed in a manner similar to free-living Turbellaria, pumping food in and waste out via the oesophagus. Gland cells surround the posterior oesophagus and secrete their contents into the lumen. These are thought to contain hydrolases which initiate the processes of digestion in the intestinal caecae. The caecal walls are composed of flattened epithelial cells which have both a secretory and absorptive function. Their luminal surfaces are drawn out into thin folds of cytoplasm. Small molecules enter the cells by passive diffusion and mediated transport, and in *Fasciola* there is a complex cycle of endocytosis. Haemoglobin is taken up, digested in phagosomes, and waste pigments are defaecated back into the lumen. Proteolytic enzymes are present in the gut lumen and digestion occurs at an acid pH.

Digeneans can also use their body surface as a route for nutrient uptake. Sugars and amino acids enter the body across the tegument by both active and passive mechanisms. Attempts have been made to determine the relative importance of the gut and the body surface in digenean nutrition, but experiments are complicated by the difficulty of persuading worms like *Schistosoma* to open their mouths and ingest test substrates *in vitro*.

Tapeworms, in contrast, lack an intestine and must rely on the body surface for uptake of nutrients. The rat tapeworm, *Hymenolepis diminuta*, has served as an experimental model for numerous studies on cestode nutrition. There are remarkable structural and functional similarities between the body surface of the tapeworm and the luminal surface of the gut. The surface area of the tegument is increased by numerous microtriches, finger-like projections with flattened tips. The tapeworm possesses membrane hydrolases as an integral part of its surface and can also adsorb host enzymes onto the membranes. These membrane enzymes digest substrates immediately prior to absorption into the tissues. A variety of sugars, amino acids, short- and long-chain fatty acids, purines and pyrimidines are actively transported across the tegument and can be accumulated in the worm tissues.

Hymenolepis also appears to respond to nutrient levels in the intestine by changing its position and degree of extension of the strobila. Its position coincides with the glucose gradient in the intestinal lumen, but the stimulus for the migration apparently is not sugar levels. The migration is thought to be a mechanism which enables the worm to compete successfully with the intestinal tissues for nutrients. These observations should put an end to the common notion that tapeworms are inert sacs soaking up nutrients. They have also been shown to possess a variety of sensory nerve endings.

Parasitic nematodes, with some exceptions, possess a well-developed alimentary tract. The body surface appears to be prevented from playing a significant role in nutrition by the inert secreted cuticle which covers it. (At least one parasitic nematode lacks a gut and absorbs nutrients across modified regions of hypodermis over which the cuticle is absent.)

Nematodes are restricted by the structure of the pharynx (oesophagus) to a diet of fluids and finely particulate material. The fluids of the nematode pseudocoelom are under a pressure against which food has to be forced into the intestine by the pumping action of the pharyngeal muscles (Fig. 5–1). The pharynx also contains gland cells which synthesize

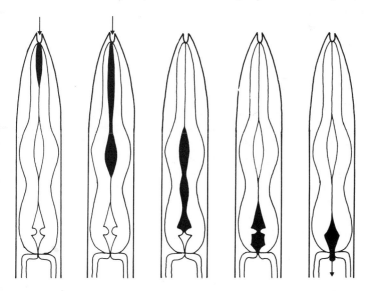

Fig. 5–1   Feeding mechanism in the nematode *Rhabditis axei* (after MAPES, C. J. 1965). *Parasitology*, 55, 583). Arrows indicate entry and exit of food from the pharynx. Dark areas show passage of food along pharynx.

and secrete hydrolases. These enzymes may be released into the buccal cavity to bring about extra-corporeal digestion of tissues (e.g. by hook-worms). The carbohydrases, lipases and proteases which have been demonstrated in the intestinal lumen may also originate in the pharynx or in the intestinal cells. The intestinal epithelium has a well developed microvillar brush border with which membrane enzymes are associated. Fluids and small organic molecules can be transported across the epithelium, though whether actively or passively is not clear.

The view has long been held that parasitism is primarily a nutritional association in which one partner benefits at the expense of the other. It is perhaps surprising, therefore, how few universal nutritional adaptations there appear to be when parasites are compared with free-living relatives. The major change exemplified most by tapeworms, to some extent by digeneans, and very little by nematodes, is the use of the body surface for uptake of nutrients with a corresponding diminution in intestinal function.

5.3   Respiration and energy metabolism

One of the principal features most likely to influence parasite respiration is the availability of oxygen. The chief problem of interpretation is to relate the results of *in vitro* studies on respiration to the situation encountered by the parasite *in vivo*. Most endoparasites can tolerate a lack of oxygen for varying periods, depending on the species. Some Protozoa (*Entamoeba*, *Trichomonas*) can be cultured successfully under virtually anaerobic conditions but will consume oxygen when it is available. Other Protozoa, flatworms and nematodes are facultative anaerobes whilst in the vertebrate host. In contrast, their free-living stages are virtually always aerobes.

The size of the parasite is important in determining whether it can obtain sufficient oxygen for its respiratory needs. In small Protozoa, the diffusion of oxygen into the tissues is unlikely to limit aerobic respiration, except at very low partial pressures, whereas in large parasites such as tapeworms or *Ascaris* it must frequently be limiting. Some parasites (e.g. *Ascaris*) can counteract this by accumulating an oxygen debt. There is a build-up of metabolic intermediates in the tissues during anaerobiosis. When oxygen becomes available its consumption is higher than normal until the excess metabolites have been used. Attempts to determine the relative importance of aerobic and anaerobic respiration as sources of energy for a parasite have been controversial. The current concensus is that, in spite of the availability of oxygen to many parasites, anaerobic respiration predominates. Whatever oxygen consumption is possible may be of real significance, and some processes such as collagen formation in *Ascaris* and egg formation in *Fasciola* have an obligate requirement for oxygen.

The presence of cytochromes in parasite mitochondria is generally related to the extent of aerobic respiration. In some species, for example the slender trypomastigotes of African trypanosomes, they may be lacking. Haemoglobins are found in some digeneans and nematodes but have not been reported from tapeworms. They may function to facilitate oxygen transport or as a tissue oxygen store. Their properties are, however, different from mammal haemoglobins, particularly in respect of the tenacity with which they hold on to bound oxygen. In digeneans the haemoglobins are concentrated in the vitellaria and around the uterus where egg formation occurs. In nematodes such as *Ascaris* there are two different forms, one in the body wall and one in the pseudocoelomic fluid.

The major energy store in most parasites is glycogen, but some Protozoa (e.g. *Eimeria*) store amylopectin (a form of starch). In most animals deprived of oxygen, the rate of utilization of glycogen increases to compensate (the Pasteur effect). This does not happen in endoparasites, with the exception of a few nematodes. In other words, there is no feedback between oxygen availability and the rate of glycolysis.

In most endoparasites the conversion of carbohydrate to pyruvate is via a typical glycolytic pathway, but from there onwards innumerable variations occur. The parasitic flagellates for example, depending on species, may release lactate, pyruvate, acetate, succinate, glycerol or carbon dioxide as end products of energy metabolism. There is now much evidence that the movement of a parasite to a new environment may be accompanied by shifts in metabolic pathways. The slender trypomastigote of African trypanosomes in the bloodstream has mitochondria lacking cristae, no tricarboxylic acid (TCA) cycle and respires glucose to pyruvate. Entry into the gut of the fly vector is accompanied by the appearance of cristae, the elaboration of TCA cycle enzymes and respiration of glucose to carbon dioxide. These changes are controlled by the kinetoplast DNA.

Studies on the intracellular stages of *Plasmodium* are more difficult to perform. It has been shown, however, that the trophozoite in the erythrocyte has mitochondria lacking cristae, no functional TCA cycle, and respires glucose to lactate. The oocyst in the mosquito, and sporozoites, have an aerobic metabolism and functional mitochondria with cristae. There is no satisfactory explanation of why *Trypanosoma* and *Plasmodium* should have adopted an anaerobic respiration in the presence of abundant oxygen. A similar situation prevails in the digenean *Schistosoma*, where lactate is the main end product, whilst in *Fasciola* acetate and propionate are predominant. In the tapeworm *Hymenolepis*, lactate and succinate are the end products. The nematode *Ascaris* goes one step further and synthesizes C_4, C_5 and C_6 fatty acids.

The pentose phosphate pathway, an alternative to glycolysis, is present in some parasites but is quantitatively unimportant as a source of energy.

The TCA cycle may often be non-functional or hard to demonstrate. Certain of the constituent enzymes may synthesize metabolites but others are lacking (e.g. in *Hymenolepis*, *Fasciola* and *Ascaris*). Mitochondria are nearly always present in the tissues of these parasites, but presumably perform functions not associated with TCA cycle enzymes.

Neutral lipids are often present in parasite tissues, particularly those of digeneans and tapeworms. In encapsulated stages and free-living larval nematodes they may serve as long-term energy stores. However, few endoparasites are able to use them to any great extent even under aerobic conditions. The tapeworm *Hymenolepis* lacks some of the enzymes necessary for the cycle of β oxidation and the nematode *Ascaris* appears to use the cycle to synthesize the volatile fatty acid wastes. In *Fasciola* the accumulated lipids pass into the protonephridial system to form one of the major excretory products. In *Hymenolepis* the lipids taken up from the host gut are incorporated into the developing eggs to form the food reserves. The same is probably true of the lipids of *Fasciola* and *Ascaris*.

Proteins may also serve as an important energy source for parasites. For example the protozoans *Trypanosoma*, *Trichomonas* and *Plasmodium* will produce ammonia when starved of sugar, indicating a switch to amino acid breakdown. Ammonia and urea are the usual end products of protein metabolism. The ammonia can have a variety of origins but the mechanism of urea formation is not clear. Helminths appear to lack the enzymes necessary for urea production via the ornithine cycle. *Ascaris* releases a variety of amines as end products of protein metabolism. Proteins are a dominant feature in the diet of tissue feeders, particularly those such as hookworms, *Fasciola*, and *Schistosoma* which consume blood. *Fasciola* has high levels of free amino acids in the tissues and their breakdown may account for the excretion of the large quantities of acetate, propionate and other volatile fatty acids.

5.4   Osmoregulation and excretion

Marine invertebrates are, to varying degrees, osmoconformers but the ionic composition of their cells differs from that of the surrounding interstitial fluids and of the seawater in which they live. Freshwater invertebrates must maintain both osmotic pressure and the concentration of inorganic ions above that of their environment. It can be anticipated that the free-living larval stages of parasites in soil (nematodes) or freshwater (e.g. miracidia) will possess mechanisms to control their fluid composition and concentration. The question is, to what extent does an endoparasitic existence modify this control?

An excretory and osmoregulatory function has been ascribed to a variety of invertebrate organs. In freshwater Protozoa, the contractile vacuole supposedly functions to expel excess water from the cell. Unfortunately the minute size of the organelle limits the scope for direct

nvestigation. A contractile vacuole is not present in *Entamoeba* but has
•een reported from some trypanosomes.

Both free-living and parasitic flatworms possess a well-developed
•rotonephridial system. In digeneans this consists of extensive
•ranched canals, bounded by a thin epithelium, ramifying through the
leeper tissues. The canals join together distally to form a reserve bladder
vhich opens to the exterior via a pore. Proximally they are intracellular
nd end in flame cells (Fig. 5–2). Each flame cell encloses a bunch of

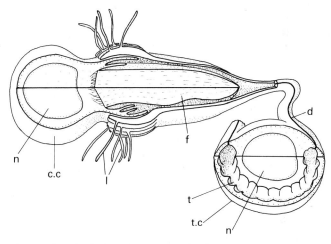

ig. 5–2 Flame cell and tubule in the protonephridial system of *Fasciola hepatica*.
c – cap cell, d – desmosome, f – 'flame', l – leptotriches, n – nucleus, t – tubule,
c – tubule cell.

agella which, by beating rapidly in unison, are thought to create a region
f reduced pressure around their base. This pressure-drop across the
arrel may be sufficient to suck interstitial fluids into the lumen of the
anal. Tapeworm protonephridia have a similar basic structure, but there
re longitudinal canals in the strobila emptying to the exterior at the
rminal proglottid, and transverse canals in each proglottid.

Nematodes possess an excretory system of very variable structure. It
ay be glandular, or part gland and part tubule. The tubules run along
e lateral hypodermal chords. A pore is generally situated on the
nterior mid-ventral surface and the tubules connect with it via a
ansverse vessel. Contractile ampullae have been reported near the pore
a the larvae of some strongyles.

Virtually nothing is known about the ability of parasitic Protozoa to
smoregulate or, if they are osmoconformers, the extent to which they
an tolerate fluctuations in osmotic pressure. All the experimental
ridence suggests that the endoparasitic stages of digeneans and

tapeworms are osmoconformers. This lack of ability to control body flui
composition, compared with free-living freshwater Turbellaria fc
example, may be related to the increased use of the body surface for th
uptake of nutrients. A body surface permeable to organic solutes is als
likely to be permeable to water. If parasitic flatworms cannc
osmoregulate, what then is the role of the well developed protc
nephridia? There is evidence that they serve as the route for the excretio
of some metabolic wastes from the deeper tissues. In tapeworms, lactat
and urea have been identified in the fluid; in digeneans, neutral lipids an
volatile fatty acids. Other wastes such as ammonia are thought to leave th
worms primarily by diffusion across the body surface.

Endoparasitic nematodes similarly show little ability to osmoregulat
and the variability of the excretory system suggests that it may serv
several purposes. If it functions in ionic and osmotic regulation, then th
is probably by secretion in the manner of an insect Malpighian tubule an
not by filtration as in protonephridia. There is evidence from studie
employing dyes that wastes are transferred from the gut cells to th
pseudocoelom and then expelled via the excretory tubules. In som
nematodes, enzymes such as acetylcholinesterase are released via th
excretory pore, supposedly causing localized inhibition of intestin:
peristalsis to prevent worm expulsion. The secretory activity of the syster
has also been implicated in the processes of moulting the cuticle.

5.5 Synthetic processes and reproduction

Although many parasites have a phenomenal capacity fc
reproduction there is only fragmentary knowledge of the physiologic:
processes associated with it. All parasitic animals can synthesi:
polysaccharides, proteins and nucleic acids but are to some exter
dependent on the host for provision of the metabolic precursors. Thu
certain amino acids are essential components of the parasite diet. Little
known about requirements for purines and pyrimidines. Trypanosom
have a limited ability to synthesize both classes whilst *Plasmodium* ca
synthesize pyrimidines but not purines. Parasitic flatworms an
nematodes readily assimilate both and incorporate them into nucle:
acids. It must therefore be assumed that they have limited powers fc
synthesis of the precursors.

The body tissues of flatworms are a solid mass from which individu:
organs cannot be isolated and, as yet, little attempt has been made t
determine the proportion of synthetic activity which is distribute
between different physiological processes. In such circumstance
synthetic processes can be studied by the incorporation of isotopicall
labelled precursors into the macromolecule under investigation. Th
distribution and translocation of the label can then be traced b
autoradiography.

Apart from reproduction, the principal synthetic activities of digeneans occur in the epithelia of the tegument and intestine, with minor contributions by the protonephridia and oesophageal glands. The pattern in tapeworms is similar, except that there is no intestinal contribution. In nematodes the equivalent of tegumental synthesis is cuticle formation by the hypodermis. It seems probable that the dominant synthetic processes of parasitic animals are those concerned with reproduction.

In the majority of parasitic amoebae and flagellates reproduction is by fission, usually binary. The Sporozoa in contrast have both asexual and sexual phases. Asexual multiplication is by schizogony, the mitotic division of the nucleus many times, before division of the cytoplasm. The switch to sexual reproduction is not understood in either *Eimeria* or *Plasmodium*. It may require a special type of schizont or a physiological trigger. In *Plasmodium* the process of microgamete formation is triggered by the drop in temperature on transfer to the gut of the insect vector. The microgametes are motile and flagellate. In *Eimeria*, the maturation of the macrogamete is accompanied by the synthesis of inclusions containing cyst wall precursors. A diploid zygote is formed at fertilization and the first subsequent division is meiotic so that the rest of the cycle is haploid.

In flatworms the morphology of spermatogenesis, oogenesis and vitellogenesis has been studied by electron microscopy, but little is known about the associated physiology. The females of *Schistosoma* will not become sexually mature in the absence of the male, implying the need for chemical or mechanical stimulation. The process of mating in schistosomes appears to involve largely tactile stimuli. The gynaecophoric canal of the male is richly supplied with sensory nerve endings. In the hermaphroditic digeneans cross-fertilization seems to be the rule, probably by direct impregnation. Tapeworm proglottids are protandrous (male first, female later). It is not clear whether fertilization is by impregnation or by release of sperm into the intestine and their chemotaxis to the female pore.

Spermatogenesis in nematodes is accompanied by reduction division. The spermatozoan is amoeboid with a complex structure and employs pseudopodia for movement. The cytoplasm of the fully formed oocyte contains a variety of inclusions which serve as food reserves or are associated with egg formation. One of the primary synthetic processes of the ovary is the formation of lipid ascarosides, components of the principal permeability barrier of the egg.

Mating in nematodes is stimulated by a pheromone. In a few species both sexes are attractive but in most it is the female which attracts the male. The pheromones are thought to be water soluble and slightly volatile, but none has yet been characterized. Copulation is stimulated by contact and the caudal region of the male is well supplied with sensory endings. The male holds the female, and spicules, if present, are inserted into the vulva. The amoeboid sperm are introduced directly into the vagina and migrate slowly up the uterus to its junction with the oviduct. This region serves as a spermatheca where the sperm may remain viable for some time.

5.6   Motility

Relatively little is known about the motility and irritability of parasitic Protozoa or the neuromuscular physiology of helminths, in spite of the fact that many drugs are believed to act on these systems (see Chapter 6). There is no reason to suppose that the motility of parasitic amoebae or flagellates differs from that of free-living species.

The sporozoites and merozoites of sporozoans possess a microtubule system lying immediately beneath the pellicle. Contraction of this system may be responsible for the rapid undulatory propulsion, but it could equally well serve as a cytoskeleton.

The musculature and nervous system of the parasitic flatworms is probably as well developed as that of free-living relatives. The thin, elongate muscle fibres are principally of the smooth type. Their basic arrangement, which has many variants, is a meshwork of outer circular and inner longitudinal fibres lying immediately beneath the tegument. The fluids of the body contained within the mesh serve as a hydrostatic skeleton, capable of deformation but not compression. The muscles act on the skeleton to generate alternate waves of contraction and relaxation, propagated along the flatworm body and employed for locomotion. Muscle fibres also run from the dorsal to the ventral surface to maintain the 'flat' configuration of the body. The nervous system consists of central ganglia in the oesophageal region (or in the scolex) with longitudinal motor and sensory nerves. Few investigations of the electrophysiology of the neuromuscular system have been reported. Ultrastructural studies have shown that the muscles are innervated at simple synapses. At least three different morphological types of transmitter vesicle have been identified in neurons. Acetylcholine is thought to inhibit muscle activity, serotonin to excite it, and an adrenaline-like transmitter may be present.

There appears to be little difference between the neuromuscular systems of parasitic and free-living nematodes. The muscles of the body wall consist of a single layer of longitudinally orientated cells, each of which is divided into a contractile and a cytoplasmic region. The muscles are obliquely striated but, unlike vertebrate striated fibres, contract only slowly. They are also unusual in making contact with motor nerves via elongate cytoplasmic processes. In the absence of circular muscles, the longitudinal fibres act against the combined effects of the cuticle and pseudocoelomic fluid, producing the high internal turgor pressure mentioned earlier. The cuticle contains inelastic spiral fibres, arranged in trellis formation to resist the contraction of the underlying muscles. In nematodes the undulatory propulsion of the body is generated by the passage of alternating waves of contraction and relaxation along the muscles. Unlike flatworms, the waves in the dorsolateral fibres are out of phase with those in the ventrolaterals, producing the characteristic sinusoidal body form. The nematode nervous system consists of a primary centre, the circumpharyngeal commissure, and a secondary ganglion in the rectal region. A number of longitudinal nerves, embedded in the hypodermis, are thought to relay sensory and motor information. A separate pharyngeal nerve net controls the processes of feeding. Acetylcholine functions as an excitatory transmitter and there is some evidence for serotonin and adrenaline as additional neurotransmitters.

Our fragmentary knowledge of neuromuscular function, together with the existence of sensory receptors, suggests that contrary to popular belief, the endoparasitic stages of helminths possess well-developed receptor and effector systems. For what purpose are these systems used? One obvious role is in site finding. Many helminths undergo complex migrations through the body to the site of parasitization and they must require information to terminate the migration when they eventually reach the preferred site. One example will give some idea of the complexity. The metacercaria of *Fasciola* excysts in the lumen of the small intestine. It does not reach the site of parasitization by migrating up the common bile duct. Instead it penetrates into the peritoneum. Eventually it locates the liver capsule and tunnels through the parenchyma to reach a bile duct where maturation ensues. Nothing is known of the cues which operate during this migration.

6 Chemotherapy

6.1 The value of drugs in disease control

The possession of an efficient, non-toxic drug would appear to provide an immediate answer to the problems posed by a parasitic disease. Dose the infected individual (man, or domestic animal), effect a cure and the problems are solved. For a number of reasons this viewpoint is too simplistic. It is certainly possible to alleviate or eliminate the effects of parasitism on the individual, for example, of anaemia caused by hookworms or malnutrition caused by various intestinal worms. In other instances such as elephantiasis or river blindness, the ravages of the parasite may be irreversible.

Drugs are seldom completely efficient. A single dose frequently fails to kill all the parasites harboured by the host and repeated doses may therefore be necessary. The biochemistry of many parasites is remarkably similar to that of the host. As a result, it has proved difficult to find chemical agents which will kill the parasite but produce no side effects. It may be necessary to administer the drug to human patients in hospital, where side effects such as nausea, vomiting, fever and pain can be treated. (In an outbreak of human fasciolisasis in the U.K., it was recorded that some people preferred the discomfort caused by the parasite to the greater discomfort caused by drug treatment.)

At the level of the host population the situation is complicated by the dynamic nature of parasitic diseases (see Chapter 4). In a situation of continuing transmission, the administration of a drug to a few parasitized individuals will have little effect on the total parasite population and the treated individuals will have a high probability of reinfection. Dosing of the entire population may have dramatic initial results with a rapid reduction in the total worm burden, but if other stages in the life cycle (e.g. in vectors) are not attacked, then the parasite population may recover at least to original levels. The alternative approach, employed by farmers to control liver fluke or trichostrongyle nematodes, is routine dosing of the entire host population. The total parasite population is kept at low levels and its effects minimized. This represents an artificially low equilibrium in which the drug has become simply another cause of parasite mortality, along with senescence and immune killing. The danger is that when a low parasite burden has been achieved, the dosing will be stopped. If this happens, and no other factors have changed there will be a recovery by the parasite to original disease levels. Worse can happen if the parasite normally provoked a protective immune response in the host, but the drug administration has prevented this from building up. The parasite may then spread unchecked through the susceptible host population causing more harm than if the population had never been treated. Eradication of a parasite population by drugs alone is difficult to achieve unless the host population can be protected from reinfection by prophylactic treatment.

6.2   The discovery and development of drugs

The search for new antiparasite drugs is an expensive process and for this reason they generally originate in the research laboratories of pharmaceutical companies or military organizations. Development costs are reflected in the retail price of the drug. The economics of drug treatment in livestock is relatively easy to quantify. The cost of the drugs administered and the benefits they bring in terms of increased milk, meat or wool production can be computed as some fraction of the market price of the animal. A drug will not be used if the cost of dosing outweighs the economic benefits. It is obviously impossible to use this approach for human diseases. The majority of the parasitic diseases of man are prevalent in those poorer countries least able to afford the expense of drug development. A single dose may cost as much as the total health care budget per head of population per year.

The logical starting point for the discovery of a new drug would seem to be a knowledge of parasite biochemistry and physiology. If a parasite possesses some unique metabolic quirk then it should be possible to design a chemical inhibitor which will kill the parasite without affecting the host. In practice, drugs are not developed in this way and determining the mode of action is often the final, not the initial, step in the sequence of events.

The first step is the establishment in the laboratory of a suitable model of the disease. For economic reasons the laboratory host is usually a rodent. The parasite may be the same one against which the desired drug will eventually be used, or a related species. *Schistosoma mansoni* will develop in the white mouse and this serves as a suitable laboratory model for schistosomiasis. *Plasmodium berghei*, a species of rodent malaria, is used as the model for human malaria and two rodent nematodes *Nippostrongylus* and *Nematospiroides* are used as models for hookworm, and the trichostrongyles of grazing animals respectively. No suitable model is available for some diseases. For example, the rodent filarial nematodes *Litosomoides* and *Dipetalonema* do not parasitize the same tissues as human filariae. *In vitro* screening of compounds has often been advocated to reduce the number of animals used for experimentation. Unfortunately many compounds active *in vivo* are inactive *in vitro* and *vice versa*.

Once the laboratory model has been selected, a large number of compounds can be tested for anti-parasite activity. (In a recent programme by the U.S. Army 250 000 compounds were screened against *Plasmodium*.) A compound is administered to infected hosts and after a suitable interval its effect on the parasite burden is assessed. If the drug is to function as a prophylactic, for example a sporonticide (sporozoite killer) against malaria, then it may be administered before infection of the host.

When promising compounds have been identified, they are put through a second and more rigorous screening procedure. The concentrations required to give 50% and 90% parasite kill (effective doses: ED_{50} and ED_{90}) are determined. Different routes of administration are tried such as by mouth, intra-peritoneally or subcutaneously. The spectrum of activity against different parasite stages and, most important, the toxicity of the compound to the host are determined. Toxicity is conventionally referred to as the LD_{50}, the dose which is lethal to 50% of the hosts.

Compounds which emerge unscathed from the secondary screen may, if they are to be used on the human population, require a further phase of testing in a laboratory model more closely related to the human disease. In malaria, for example, this is done with *Plasmodium cynomolgi* in rhesus monkeys or with *P. falciparum* in the owl monkey *Aotus*. The final phase involves clinical studies and field trials. The potential drug is administered to uninfected hosts (human or animal) and its effects on physiology and biochemistry are noted. A compound which passes this test may then be tried in the field on a selected group of infected hosts. If the drug is effective against the parasite but does not unduly harm the host, it may then be subjected to large scale trials before being marketed world-wide. The whole process from start to finish will have taken five years or more.

6.3 Toxicity and mutagenicity

The biochemical and physiological processes of parasites are so similar to those of the host that there is little opportunity to exploit any differential toxicity to a drug. Many early anti-parasite agents were compounds containing arsenic or antimony and were therefore very toxic to the host. The list of side effects associated with even the safest drug is usually long in a proportion of patients to whom it is administered. However, when the alternative is a lingering death from the disease, this may be entirely acceptable. No amount of testing in animals can make a drug totally safe for humans. A number of anti-schistosome compounds have been found to cause nervous disorders in some patients. The reason was eventually identified as a failure by affected persons to metabolize the drug. Repeated administration caused a rise in concentration in the bloodstream to critical levels. Another example is the drug Primaquine used against malaria. It can cause haemolysis of erythrocytes.

The problems of mutagenicity and teratogenicity (production of mutations and developmental defects respectively) have only been recognized relatively recently. Evidence was accumulated over a number of years that an effective drug widely used against schistosomiasis was a potent mutagen. It caused carcinoma of the liver hepatocytes in a proportion of infected mice. Some teratogenic effects were also noted when the drug was given to pregnant mice. It has no effect on normal mice and it appears that the schistosome infection predisposes the liver cells to carcinoma. These findings have proved controversial and the drug is still widely used in many countries partly because it is so effective and partly because a connection with hepato-carcinoma in man has not been demonstrated.

6.4 Mode of action

It is difficult to identify the precise mode of action of a drug. It may have a single direct effect at some point in metabolism, but almost certainly there will be a variety of secondary effects. Alternatively the drug may act at several unrelated points. The task of the experimenter is to unravel the cause of the often complex range of responses occurring in the parasite. Many thousands of compounds can be classed as anti-parasite drugs and very few have a common mode of action. For this reason the most profitable approach is to look at the reactions of a parasite species to different drugs in the hope that particularly sensitive points in metabolism can be identified.

The malaria parasite *Plasmodium* has several such key points. Detailed studies of the drug Chloroquine suggest that it binds to receptors on the parasite surface and is taken into the cell in a pinocytotic vesicle. It is thought to raise the pH of the vesicle contents, depressing the acid digestion processes and leading to starvation of the parasite. Another group of anti-malaria drugs appears to starve the parasite by acting on the membranes of the host erythrocytes to alter their permeability to nutrients such as glucose. The pathway used by *Plasmodium* to convert p-aminobenzoic acid to folic acid and folinic acid is the target for other drugs. Folic acid and its derivatives are important cofactors in purine and pyrimidine synthesis so that the secondary effects are manifested in nucleic acid metabolism. A fourth group of anti-malarial agents are thought to act on the mitochondria of the trophozoite, interfering with energy metabolism.

A number of anti-trypanosome drugs are concentrated within cytoplasmic vesicles in a way similar to Chloroquine in malaria, but whether they interfere with cell nutrition is unknown. Several glycolytic enzymes in trypanosomes are inhibited by organic arsenical compounds. The cytoplasmic DNA of the kinetoplast also seems particularly susceptible to drugs.

The mode of action of drugs on helminths is even less well understood. The commonest macroscopic effect is paralysis of the musculature. If the worm is a gut parasite, even a transient loss of motility can lead to it expulsion by intestinal peristalsis. In schistosomes, however, the paralysi results in the worms passing down the hepatic portal vein (the hepatic shift) but they get no further than the liver sinuses. Unless drug treatmen is prolonged or the drug is persistent, the schistosomes may recover and migrate back up to the mesenteric vessels. There is little evidence that anthelmintics act directly on the worm muscles to cause the paralysis. A few compounds interfere with neurotransmission by inhibiting the enzyme acetylcholine esterase, but most seem to function by inhibiting specific enzymes, especially those concerned with energy production. Organic antimonials inhibit the glycolytic enzymes of schistosomes. Drugs such as Tetramisole and Thiabendazole inhibit the succinic dehydrogenase and fumarate reductase of nematodes.

The limitations of *in vitro* testing are illustrated by the effect of cyanine dyes on schistosomes. They will markedly depress the oxygen consumption of the worms *in vitro* but exert no harmful effects *in vivo*. This is presumably because the worms are not heavily dependent on oxidative metabolism as an energy source.

6.5 Drug resistance

The repeated use of a drug can act as a selection pressure on the parasite, causing the evolution of resistant strains. This resistance may be stable or unstable, i.e. it may become a permanent feature of the parasite's metabolism, or the parasite may revert to a sensitive form after cessation of treatment. Drug resistance poses the greatest problems with protozoan diseases such as malaria or coccidiosis, presumably because their life cycles are sufficiently short to permit a rapid evolution. However, drug resistance by flatworms and nematodes has also been reported. The common practice among farmers of changing their brand of anthelmintic every few years is probably an instinctive response to the development of resistance by the parasitic helminths of livestock.

Plasmodium falciparum has become resistant to the drug Chloroquine in South America and South East Asia. This resistance has been correlated with a reduced number of membrane-binding sites for the drug. The Chloroquine-resistant trophozoites also appear to switch on their aerobic metabolism prematurely whilst still in the erythrocyte.

Resistance of trypanosomes, in both man and domestic livestock, to a number of drugs has been reported. Anti-trypanosome drugs have been classified according to their ionic characteristics at blood pH, using the phenomenon of cross-resistance. It is possible that in resistant strains the ionic character of the drug receptor has changed in such a way that drug binding is limited. The drug resistance of trypanosomes appears to originate without sexual reproduction and genetic recombination. It presumably arises by spontaneous mutation and spreads rapidly by the selection of resistant individuals.

The arbitrarily selected examples given above illustrate the limited capability of the currently available drugs. They suggest a continued need to search for effective, safe, single-dose formulations against virtually all protozoan and helminth parasites.

7 Host-Parasite Interactions

7.1 Immunity and disease

Each partner in the host-parasite relationship is greatly influenced by the activities of the other. The parasite interferes with host physiology to produce malfunctions which are collectively called *disease*. The host can use the complex mechanisms of the immune response to combat the parasite. The interlocking themes of immunity and disease are the subject of this chapter. The success of the immune response in controlling specific parasites is examined, together with the prospects for the production of vaccines. The vexed question of host specificity or innate resistance is not dealt with here. Some parasites (*Fasciola*, *Trichinella*) will develop in numerous unrelated hosts; others (*Wuchereria*) in only one. Much has been written on this subject but our knowledge remains largely speculative.

The immune response is conventionally represented as being mediated by humoral factors, the immunoglobulins, or by sensitized cells of the lymphoid system. In reality the response is far more complex and immunity to the parasite, if it occurs at all, is the result of interdependent humoral and cellular processes.

7.2 Components of the immune response

Lymphocytes, derived from stem cells in the bone marrow, play a central role in the immune response. They are heterogeneous in their properties and several sub-populations have been identified. One major group, the 'T' cells, matures under the influence of the thymus and transforms into sub-populations such as helper cells and killer cells. The other major group, the 'B' cells, are conditioned in birds by the Bursa of Fabricius, an organ near the cloaca. Mammals also possess a population of B cells but a similar conditioning tissue has not been identified. The B lymphocytes give rise to plasma cells whose primary function is the synthesis of immunoglobulins (Ig's). Both B and T cells are characterized by the possession of surface receptors which will bind foreign molecules.

Other cells also play an important role in immunity. These include macrophages, both fixed and circulating, and granulocytes (polymorphonuclear leucocytes). In addition to the immunoglobulins there are other humoral factors, collectively called the complement system, which circulate in the plasma. Complement is frequently a necessary adjunct to the action of antibodies, particularly in the lysis of target cells.

7.2.1   B cells and the production of antibody

The molecules of parasite origin which stimulate the host's immune response are termed antigens. These antigens may bind directly to the surface of B cells to stimulate them or their progeny to produce specific antibodies. Alternatively, B cells may require the assistance of helper T cells and of macrophages to respond to some antigens. The nature of the help given is problematic. The antigen may form an intercellular bridge between the T cell and the B cell which stimulates the latter to produce antibodies. A more likely explanation is that the helper T cell is stimulated by antigen to shed its receptors, referred to as IgT. These are bound at the surface of an accessory cell, the macrophage, and there interact with B cells to stimulate antibody production.

A number of different classes and subclasses of antibody or immunoglobulin have been identified, of which IgA, IgE, IgG and IgM are the most important. IgG is the common circulating antibody, IgA can be secreted across epithelial surfaces, and IgE is associated with cells in immediate hypersensitivity reactions. They are proteins which differ in their detailed molecular structure but are basically Y-shaped molecules. There are two binding sites for antigen on each molecule, one at the tip of each arm of the Y. The IgM molecule consists of five Y-shaped sub-units joined by their tails.

7.2.2   Immune effector mechanisms

The immune system exhibits a whole range of reactions against parasites. These may involve antibodies alone, cells alone, or some combination of the two.

(1) *Neutralization by antibody.* All classes of antibody may exert their effect by binding to and neutralizing either soluble antigens or antigens which form part of the parasite structure.

(2) *Complement-mediated cell lysis.* The binding of one molecule of IgM or of two molecules of some classes of IgG to target cells, may initiate the sequence of complement activation or fixation. This damages the cell membrane and causes lysis. Molecules capable of attracting polymorphonuclear leucocytes are also produced during complement fixation, presumably to facilitate ingestion of the debris resulting from lysis of the target cells.

(3) *Antibody-stimulated phagocytosis by macrophages.* The phagocytosis of pathogens is greatly enhanced if they become coated with cytophilic antibodies, also called opsonins. The adherence of the pathogen to the macrophage may require the presence of complement. The macrophage is itself activated by the release of a humural agent, macrophage arming factor, from T cells which have been sensitized by specific antigen derived from the pathogen.

(4) *Antibody-mediated cell lysis by lymphocytes.* The binding of IgG to a

variety of target cells will stimulate normal unsensitized lymphocytes, which can be derived from a non-immune donor host, to kill target cells. This system is not dependent on complement or macrophages, and the lymphocytes are not T cells or B cells. They are generally referred to as 'K' cells.

(5) *Immediate hypersensitivity*. This mechanism involves the cooperative action of IgE and mast cells. The latter are found in many epithelial surfaces such as the skin, lungs and intestine. IgE, which is termed a reaginic or homocytotropic antibody, is present in the circulation in very low concentrations. The bulk is bound to receptors on the surface of mast cells, which are then said to be sensitized. When the specific antigen which elicited IgE production binds to the antibody on the sensitized cell it causes the release of granules. These contain pharmacologically active compounds such as histamine, serotonin, kinins, eosinophil chemotactic factor, etc. The compounds react on the surrounding tissues, causing smooth muscle contraction and increased permeability of blood vessels to bring about a local anaphylactic reaction. The reaction appears very rapidly after the injection of antigen into a sensitized animal, and for this reason is called an immediate hypersensitivity response.

(6) *Delayed hypersensitivity*. In the immune reaction to tumours and tissue grafts, circulating antibodies may not be detectable and the response is predominantly mediated by T cells. The resulting inflammatory reaction develops only slowly after the injection of specific antigen into a sensitized animal, reaching a peak 24–48 hours later. The T cells have receptors specific for the particular antigen. Upon interaction with the antigen they are transformed to undergo division and also release a whole variety of chemical factors, collectively called lymphokines (e.g. macrophage chemotactic factor). The sensitized lymphocytes bind to and exert a cytotoxic effect (killer function) on target cells.

7.2.3   *Regulation of the immune response*

The mechanisms by which the immune response is regulated are complex and imperfectly understood. However, there would be an obvious advantage to be gained by a parasite which interfered with these mechanisms in order to diminish the response directed against it. A parasite may be able to induce *tolerance* in the host by exposing the T and B cells to massive amounts of antigen. After a time the host would no longer regard the parasite as foreign. Parasites may also cause *suppression* of T cell formation.

Another possibility is that parasite antigens may interfere with the binding of IgT molecules, derived from T cells, to macrophages. Such *antigenic competition* would depress the production by B cells of those antibodies which require the participation of helper T cells and macrophages.

7.3   The immune response to parasites

Many of the parasites described in this book are able to live within the verebrate host for a period of months or even years. This fact cannot easily be reconciled with the development of protective immunity on the part of the host. Indeed, the harbouring of large numbers of parasites, and the ease with which superinfection or reinfection can occur, suggest that in some cases protective immunity may be virtually non-existent. There are many possible explanations for this state of affairs. The similarities of host and parasite biochemistry and physiology might result in parasites being only weakly antigenic. Alternatively, the parasite could evoke an immune response whilst remaining safely beyond the reach of humoral or cellular effectors, in the lumen of an organ such as the gut. A primary immune response takes about two weeks to develop. Thus, if the host responds to a parasite during its tissue migration phase, then the parasite might well have reached a safe site before the response becomes effective. However, it is obvious that explanations such as these cannot be applied to parasites which inhabit the bloodstream, a potentially hostile environment.

7.3.1   Protozoa

Protozoa can be found parasitizing virtually every habitat which the vertebrate body can provide and we might therefore anticipate that the immune response directed against them will take a variety of forms. The trophozoites of *Entamoeba* in the intestine frequently stimulate the production of high levels of antibody circulating in the bloodstream, but this has little effect on the parasite. However, trophozoites reaching the liver can confer on the host a lasting immunity to further liver abscesses, although the mechanism is not understood.

In contrast, the coccidia which also inhabit the gut can elicit a more potent response, probably because they are intracellular parasites. In *Eimeria* the extent of the host response varies with the species of parasite and there is little cross-reaction. Following infection, antibodies can be detected in the bloodstream. These may be opsonins which bind to sporozoites and stimulate their phagocytosis by macrophages. Plasma cells in the gut wall may also produce IgA secretory antibodies which can pass into the gut lumen. In spite of the antibody production, attempts to transfer immunity to naïve hosts using immune serum have not been very successful. Infection with *Eimeria* results in a massive cellular infiltration of the intestinal epithelium by plasma cells, lymphocytes and macrophages. It is probable that these cells are the major source of immunity.

The discovery of the intestinal phase of *Toxoplasma* in cats is too recent for much to have been learned about the immune reaction to it. The tissue phases of *Toxoplasma* have been more intensively investigated. After infection *Toxoplasma* spreads through the body via the lymphatics and bloodstream, invading tissues and destroying cells. This elicits the appearance of both antibodies and sensitized T lymphocytes, which cause a reduction in the numbers of parasites. These may, however, persist in the tissues for many years and toxoplasmosis can still be transmitted congenitally to the offspring. In some sites such as the brain and eye, the parasite may continue to proliferate, ostensibly because the immune response cannot reach it. The relative protective roles of antibodies and cells are not clear. However, transfer of sensitized lymphocytes to a naïve host confers on it more protection than does transfer of antibodies, suggesting that the cellular response is the more important. Sensitized T cells and activated macrophages appear to be the major cytotoxic agent. In some hosts *Toxoplasma* can suppress the immune response. In mice the suppression results from antigen competition depressing antibody production. Paradoxically, animals immune to *Toxoplasma* are also resistant to infection by a number of bacteria.

The species of flagellate protozoans grouped within the genus *Leishmania* present us with an example of tissue parasites which have a very complex interaction with the host immune response. An initial skin infection results in a sore or ulcer containing macrophages harbouring the amastigote form of the parasite. This lesion develops over several months and may cure spontaneously or the parasites may be transported to secondary sites in the skin and viscera. The host's reaction to the lesion is predominantly cellular, but humoral antibodies may play some part. (It has been suggested that these antibodies are active mainly against promastigote forms injected by the sand-fly intermediate host.) During development of the lesion there is a build up of T lymphocytes sensitized against leishmanial antigens. Immediately before self-cure the lesion is invaded by these T cells which may exert their cytotoxic effects on the parasite or on the infected host cells either directly or via activated macrophages. Following self-cure, immunity to reinfection is solid and persistent.

Unfortunately, in many individuals the disease does not progress to self-cure. Leishmaniasis has been described as a spectrum of disease. At one end there is little host response and abundant parasites cause extensive lesions. The localized self-healing lesions represent the intermediate state, and at the other end of the spectrum there is a massive cellular response. Few parasites can be detected but the response also destroys much host tissue. Where this exaggerated immune response occurs, the harmful allergic effects appear to have been dissociated from the beneficial protection it would normally confer. (The mechanism may involve desensitization of T cells.) At the other extreme, the lack of immunity may be accompanied by high antibody levels, but delayed hypersensitivity fails to develop. This failure may be caused by the induction of selective tolerance in T lymphocytes. A further side effect of the visceral form of leishmaniasis may be an anaemia, the result of erythrocytes having a shortened life span. It is thought that the parasite induces the host to produce auto-antibodies against the erythrocytes.

The trypanosomes are parasites of the bloodstream and might be expected to provoke a classic immune response leading to their elimination. In some respects this is true of infection with *T. cruzi*, but infection with African species produces a very different result.

In man, infection with *T. cruzi* is followed by rapid parasite multiplication, and the number of parasites detectable in the bloodstream reaches a peak 3–4 weeks later. Thereafter, there is a decline due to the onset of protective immunity. Protection is not absolute, since low numbers of parasites may be harboured, perhaps for life. The mechanism of immunity seems to be predominantly humoral, and immunity can be passively transferred to a naïve host by injection of immune serum. Infection is followed by a rise in specific IgG and IgM levels in the blood which reach a peak around the time that the parasites are disappearing. The antibodies may be opsonins stimulating phagocytosis of the trypanosomes. The complement system is probably also involved in direct lysis of trypanosomes after antibody binding.

The pattern of an initial infection with the three most important African trypanosomes, *T. brucei*, *T. gambiense* and *T. rhodesiense*, is very similar to that in *T. cruzi*. However, following the first decline in numbers there is a resurgence of the parasite and then a further decline. This pattern is repeated until the later stages of the disease when the nervous system is invaded. Untreated infections of the African trypanosomes are usually fatal (after a few months in *T. rhodesiense*). The decline in trypanosome numbers results from the appearance of specific IgG and IgM antibodies in the blood which cause lysis by complement fixation, and phagocytosis.

Two main reasons have been suggested for the continual resurgence of the trypanosomes: antigenic variation, and suppression of the immune response. There is now ample evidence that antigenic variation occurs. A different specific antibody appears in the bloodstream each time the parasite population reaches a peak and each wave is unaffected by the antibody to the wave which preceded it. The parasite is apparently able to synthesize an endless sequence (at least 22) of variant antigens against which the host responds by making the appropriate antibody. There is no evidence that an antigenic variant can occur twice in an infection, and the order of appearance can differ. The antigens have been identified as glycoproteins and they are present in the surface coat of the trypanosome, external to the plasmamembrane.

The way in which the trypanosome generates this antigenic diversity has not been explained. It seems ironic that one of the outstanding questions in immunology today is how the diversity of antibodies is generated by the immune system. It appears that the African trypanosomes have outwitted their hosts by employing an analogous mechanism to generate the diversity of antigens.

The other important mechanism which permits trypanosomes to survive is suppression of the immune response. Initially this appears to be the result of inhibition of helper T cell function by antigenic competition, causing a reduction in IgG production by plasma cells. Later in the infection IgM production is reduced and there may be direct damage to both T and B lymphocytes.

The pattern of immunity which develops to infections with malaria is even more complex than that to trypanosomes. The various species of *Plasmodium* can persist in man for a period of years and the parasites obviously possess mechanisms for mitigating the effects of the immune response directed against them. In endemic regions malaria is a disease of children, who show repeated and severe attacks. These are thought to occur as a child encounters different strains of the parasite. By early adulthood the level of parasites in the blood is low, and the individual is virtually immune to further attacks. It has been suggested that the prolonged period of time required for development of immunity is the time taken to encounter all the different strains in a particular locality. (N.B. Superinfection with more than one strain can of course occur.)

The major responses of the host to malaria are associated with the stages in the erythrocyte, although the sporozoite and exoerythrocytic parasite may also elicit antibody production. The different strains of malaria can be characterized by the antigens which they express. However, as well as these differences, a given strain also has a repertoire of characteristic intrastrain variation. Following a primary infection with one strain, there is a rise in the number of parasites in the bloodstream and then a decline which coincides with the appearance of variant-specific antibody. (There are genetic differences between individual people which determine how they cope with this initial infection.) Resurgence of the parasite follows the decline as a new antigenic variant is expressed, and the host responds with a new specific antibody. Eventually the parasite is controlled because a degree of strain-specific immunity is built up.

There is no evidence for sensitized T lymphocytes acting directly on the parasites or on infected erythrocytes, and immunity must be mediated largely by antibodies. Two types of antibody have been described. The first does not protect the host and is specific for the infecting strain. It induces the change in antigen expressed by the parasite at the surface of the erythrocyte. Virtually nothing is known about the nature of these variant antigens or what molecular events are involved in the switch to a new variant. The concensus of opinion is that the process is an adaptation by the individual parasite and not selection of variants by the immune response (cf. trypanosomes). The second type of antibody which brings about the death of the parasite is an opsonin, capable of binding to the surface of the infected erythrocyte and to the free merozoite, thereby preventing the infection of further erythrocytes. The opsonized cells and parasites can then be phagocytozed, particularly by macrophages in organs such as the spleen. The relative rate of production of the two types of antibody determines the number of parasites in the bloodstream. If the variant-inducing antibody is produced in advance of the parasite-killing antibody, then parasite levels in the blood tend to be high.

Non-specific activation of macrophages, as evidenced by their ability to remove carbon particles from the blood, plays a secondary role in controlling the parasites. The production of protective antibodies by B cells is largely dependent on the sensitization of helper T cells by parasite antigens. In view of this, it is not surprising that suppression of the immune response to *Plasmodium* by antigenic competition has been reported.

A number of harmful side effects of the immune response to malaria can occur. In infections with *P. malariae* antigen-antibody complexes become sequestered in the basal lamellae of the glomeruli in the kidneys, impairing their function. Another damaging feature is haemolytic anaemia which results from the destruction of normal uninfected erythrocytes by phagocytosis.

7.3.2 *Platyhelminthes*

The lumen of the vertebrate gut is separated from the bloodstream by tissue barriers which might be expected to impede the action of the immune response to parasites dwelling there. Mature strobilate tapeworms lacking a rostellum of hooks (e.g. *Hymenolepis diminuta*) ought to be particularly favoured since they are unlikely to tear host tissues. However, evidence is accumulating that even these lumen-dwelling parasites are not safe from attack. Infection with some species of tapeworm results in the appearance of antibodies in the bloodstream which have no apparent effect on the established infection. Yet, if the immune response is artificially suppressed, the worms grow abnormally large. When mice are infected with the rat tapeworm *Hymenolepis diminuta*, the worms are expelled from the gut before they reach maturity, as a result of attack by the immune response. There is no evidence for a similar expulsion by the normal rat host.

Larval tapeworms developing in mammalian tissues provoke an immune response which protects the host against subsequent infections. The early tissue stages, particularly hexacanths, seem to be the most effective in eliciting this response and the contents of their penetration glands may be highly antigenic. During its development the cysticercus larva is susceptible to the immune response but there is little evidence that the susceptibility is retained after maturation. In immune hosts a secondary infection of hexacanths is eliminated and the mechanism is in part T-cell dependent. The immune response against the larval stages of tapeworms such as *Taenia* and *Echinococcus* may act as a density-dependent regulator of the parasite populations (see Chapter 4). A continuing invasion of mammal intermediate host tissues by the larval tapeworms provokes a stronger immune response which causes a more than proportional mortality of the larvae. This limits the numbers of cysticerci available to infect the final host.

The majority of digeneans, like adult tapeworms, are parasites of the vertebrate gut and virtually nothing is known about host responses to them. Some have only a brief parasitic phase and may conceivably have adopted a strategy of reproducing rapidly in the latent period before the host can mount a primary immune response. In contrast much more is known about host responses to *Fasciola hepatica* and the various species of *Schistosoma*. These blood-feeding digeneans parasitize somewhat atypical sites, the bile ducts of the liver, and the bloodstream respectively.

Infection with moderate numbers of *Fasciola* can confer some degree of resistance. However, the mature worms in the bile duct, which provoked this immunity, appear to be unharmed. The extent of resistance to reinfection depends very much on the host species, being much greater in cattle than in sheep. Rats, mice and rabbits have been used as laboratory models to study acquired immunity to *Fasciola*. In rats, the immunity which develops can be passively transferred to uninfected recipients, using either immune serum or lymphoid cells. It has been suggested that humoral immunity is effective against the peritoneal phase of migration, and cellular immunity against the hepatic phase.

The relationship between schistosomes and their human host is complex. As with *Plasmodium* there appears to be a balance between parasitization and immunity. Levels of infection are highest in children and decline steadily in adults. This pattern of disease may in part be due to the development of an immune response, and the term 'concomitant immunity' has been used to describe a situation where the host harbours mature worms but is partially resistant to reinfection.

A variety of laboratory models using mice, hamsters, rats and rhesus monkeys, have been used to investigate immunity to schistosomes. Rats and rhesus monkeys can build up sufficient immunity to eliminate virtually the entire worm population but mice and hamsters cannot. Circulating antibodies can be detected in the sera of several host species. These antibodies, together with complement, will kill worms *in vitro* by lysing the tegument but are insufficient to kill worms *in vivo*. In hosts with concomitant immunity there is a strong cellular response to a secondary infection of worms invading the skin and lymphocytes, macrophages, neutrophils and eosinophils surround the schistosomula. *In vitro* these same cells are capable of killing young worms in the presence of immune serum. Mast cells in the skin, sensitized with IgE antibodies against worms, may also play a role in killing invading schistosomula.

Three concepts have been developed to explain how mature schistosomes in the bloodstream can evade the immune response which they provoke. These concepts are not mutually exclusive and each involves a mechanism by which the antigenicity of the worm's surface might be disguised. The first is based on a form of molecular mimicry in which the host may mistake the parasite for 'self'. Antigens of parasite origin, identical to those of host tissues, are exposed on the parasite surface. The second concept is based on the demonstration that worms can acquire a coating of host molecules, called host antigens, on their surface. These molecules are glycolipids with blood group activity and it could be said that the worm is disguising itself as a very large erythrocyte. The third concept is based on the process of surface turnover. The schistosome tegument is bounded by a normal plasmamembrane over which a trilaminate layer, the membranocalyx, is secreted. This secretion has a high lipid content and may be antigenically relatively inert. The layer turns over rapidly and is sloughed off into the environment. Thus any adhering antibodies or cells would be rapidly discarded. It is probably very significant that invading schistosomula only acquire this protective secretion several hours after penetration into the skin, and so may briefly be vulnerable to the immune response.

The major pathological effect of schistosomiasis is the product of a delayed hypersensitivity immune response to schistosome eggs. During their passage through the tissues the eggs release secretions which are antigenic. Over a period of time host T cells become sensitized and when eggs accidentally become lodged in the liver or other organs, they attract these sensitized cells. The T cells in turn release lymphokines which cause macrophages and eosinophils to collect around the egg. A chronic inflammation results and some host tissue is destroyed. Finally, the egg is walled off in fibrous scar tissue called a granuloma. Liver circulation is progressively impeded, causing increased blood pressure in the hepatic portal vein. A collateral (by-pass) circulation develops, and death results from the rupture of an enlarged blood vessel. Suppression of T cell sensitization by removal of the thymus might be expected to ameliorate this pathology. Instead, the life of the host is shortened because the eggs then cause extensive necrotic damage to the liver.

7.3.3   Nematodes

Most of the important parasitic nematodes either inhabit the tissues of their vertebrate host or have a tissue migration phase in the life cycle. They generally provoke an immune response, the effectiveness of which depends on the parasite species. Complete protection against reinfection may be conferred or at the other extreme, no protection at all. Two common features of the host response to nematodes are high levels of eosinophils in the bloodstream and the production of reaginic (IgE) antibodies.

About fifty years ago it was noted that sheep infected with the trichostrongyle nematode *Haemonchus* could spontaneously eliminate the bulk of the mature worms which they harboured. Infections of other trichostrongyle species are frequently eliminated by this mechanism of 'self-cure'. The reaction is initiated when the sheep takes in a new dose of infective larvae. The host produces IgE to a primary infection and mast cells in the intestine become sensitized. However, the response does not immediately affect the worms of the primary infection which induced it. The third-stage larvae of a secondary infection release antigens when they moult. These antigens provoke a strong localized inflammation by interacting with the mast cells. This response is not the major effector of self-cure but probably alters the permeability of the intestinal epithelium, creating a 'leak lesion'. Circulating antibodies may then pass into the intestinal lumen to attack the worms. The lung worm *Dictyocaulus*, the cause of parasitic bronchitis in cattle and sheep, evokes a similar reaction from the host and a fairly solid immunity to reinfection can be induced.

The intestinal nematode of rats, *Nippostrongylus braziliensis* has been intensively studied as a laboratory model for the self-cure response. *Nippostrongylus* must be attacked both by antibody and by sensitized T lymphocytes before it is expelled from the gut. Juveniles and adults of *Trichinella spiralis* in the intestinal mucosa of a variety of hosts provoke a very similar immune response. (The trichina larvae in the muscles are of minor immunological importance.) The tissues of the intestine are infiltrated by antibody-producing cells and mast cells are sensitized by IgE. A generalized inflammation is produced at the site of parasitization but worm expulsion depends on the participation of sensitized T lymphocytes. Immunity to trichinosis in mice is transferred naturally from mother to offspring during lactation. There is also evidence that *Trichinella* can suppress the cellular immune response of its host, rendering it more susceptible to infection with other pathogens.

Partial resistance to the blood-feeding hookworms can develop in animal hosts and perhaps in man. The worm burden of a host infected with *Ancylostoma* or *Necator* may be limited by an acquired immune response. The nature of the response in man is not clear and the absence of self-cure might be due to a failure to develop effective cell-mediated immunity. There is a strong cutaneous inflammatory reaction to hookworm larvae invading hosts with an established infection, but the larvae do not appear to be damaged (cf. *Schistosoma*). A solid immunity can develop in dogs infected with the hookworm *Ancylostoma caninum*.

There is little evidence that infections with *Ascaris* in man result in immunity, but pigs are more responsive. Antibodies of several classes can be detected in the bloodstream and the migrating larvae provoke a granuloma reaction very similar to that against schistosome eggs. This implies that cell-mediated immunity may be important and the larvae act as a focus for neutrophils, eosinophils, etc. The enzymes of *Ascaris* (and many other nematodes), particularly those secreted extracorporeally, are frequently potent antigens.

A form of arrested development of larval nematodes, dependent on the immune response, was described in Chapter 3. Infection of dogs with the ascaridoid nematode *Toxocara canis* or with the hookworm *Ancylostoma caninum* provokes a reaction which stops development of the larvae but does not kill them. During pregnancy and lactation this suppression is reduced and the larvae migrate either via the placenta to the unborn young, or to the mammary glands, infecting the newborn puppies when they feed.

In comparison to intestinal nematodes, there is relatively little known about immunity to spirurids which infect the tissues and lymphatics. This is unfortunate because they are the cause of several important diseases. There are few suitable laboratory models and spirurid worms often take months to mature. Circulating antibodies can usually be detected in infected hosts but they frequently cross-react with antigens of other parasite species and appear to confer little protection, for example against the Guinea worm, *Dracunculus medinensis*. Adults of *Oncocerca volvulus* are trapped in a nodule of fibrous tissue of host origin but are apparently unharmed. *Brugia pahangi* infections in cats have been investigated as a model for the closely related *Brugia malayi* and also for *Wuchereria* in man. Cats develop resistance to *B. pahangi* after about fifty reinfections. It is accompanied by the disappearance of microfilariae from the bloodstream but the nature of the mechanism remains to be elucidated. In man infected with *Wuchereria*, microfilariae are typically present in the blood when pathological symptoms are slight, but absent where gross malformations such as elephantiasis occur. This suggests that the pathology of elephantiasis is associated with the immune response. Blockage of the lymphatics may be the result of inflammation around dead and dying worms. A small proportion of people infected with *Wuchereria* develop a syndrome called eosinophilic lung. This is characterized by very high levels of eosinophils and of IgE in the blood. Again, this reaction is the result of an over-active immune system.

7.4 Development of vaccines

Effective vaccines offer several advantages for control of parasites. They can be used to prevent disease rather than cure it and, because they utilize the host's own physiological mechanisms, they are likely to be much safer than chemotherapy. Unfortunately, work on vaccines against protozoan and helminth parasites is still in its early stages, and with a few exceptions it is impossible to say whether such vaccines will ever become a reality.

For a vaccine to have much hope of success, it is probably a prerequisite that the host develops at least a degree of acquired immunity to the parasite. From the foregoing account of immune reactions to specific parasites it will be obvious that for some, the host reaction fulfils this criterion, for others it does not. There are two main strategies for vaccine production, the use of whole killed parasites or some antigenic fraction of the same, and the use of live but attenuated (weakened) parasites. The majority of successful anti-parasite vaccines are in the latter category. The virulence of parasites can be reduced by prolonged *in vitro* culture, by chemical treatment, and by X or gamma irradiation. The practice is best suited to those situations where the infective stage is in a resistant and quiescent state, allowing the production of a vaccine with a shelf life of several weeks. Commercial vaccines have been produced in this way against two species of the lung worm *Dictyocaulus* and against the dog hookworm *Ancylostoma caninum*. Irradiated oocysts have been used to protect poultry against infection by *Eimeria*, and irradiated metacercariae have been used to protect rats against *Fasciola*. Production of vaccines against these parasites might therefore be practically, if not commercially, feasible.

For those parasites which cannot be cultured *in vitro* and do not have resistant stages, the practical difficulties are greater. A sporozoite vaccine has been produced against *Plasmodium falciparum* and tested with some success in human volunteers. It was made by the simple expedient of irradiating infective mosquitoes and then allowing them to bite the volunteers. The immunity which was induced persisted for around three months. A different approach has been tried with *Schistosoma*. Irradiated cercariae will render mice partially resistant to reinfection, but the brief life span of the cercaria rules out any practical 'live' vaccine. Attempts are therefore being made to store young schistosomula, which have been preserved in a viable but suspended metabolic state, by cryopreservation techniques similar to those used for storing sperm. If these attempts are successful, then schistosomula could be attenuated before storage, revived when required, and injected into recipients.

In at least two other parasites live vaccines have been used without the need for attenuation. Merozoites of *Plasmodium knowlesi* have been isolated in large quantities and successfully used to vaccinate monkeys against challenge infections. The other example is the deliberate infection with live virulent *Leishmania tropica* to give complete protection against subsequent infection. The ulcer produced by inoculation of the parasite may be as severe as if the disease had been contracted naturally, but it can be sited in an inconspicuous place to prevent facial scarring. The course of the infection can be controlled with drugs if required.

Vaccines against the blood-dwelling trypanosomes would be of great use in disease control. Some success has been achieved with gamma-irradiated *Trypanosoma cruzi*. However, many workers believe that a vaccine against the African trypanosomes will never be developed because of the apparently limitless capacity for antigenic variation. The outlook for vaccines against other parasites seems equally remote. In view of the success with the dog hookworm vaccine, it might be supposed that the human hookworms could be treated similarly. However, they attract a disproportionately small research effort. Our knowledge of immunity to filarial nematodes is currently too limited for assessment of the feasibility of vaccines against these parasites.

7.5   Disease

Mention has already been made (section 7.2 and Chapter 2) of the disease conditions brought about by different parasites, and some further mechanisms of disease production are considered here. The major difference between infection with Protozoa and helminths is that the latter do not multiply in the host. For this reason it is possible to harbour small numbers of helminths and yet have no symptoms of disease. Protozoa, on the other hand, can multiply rapidly from small initial numbers to eventually overwhelm the host.

The most obvious effect of harbouring intestinal parasites is the competition for nutrients and the consequent malnutrition. Many intestinal parasites may also feed on blood, causing anaemia. Another source of harm to the host is the damage to intestinal tissues (e.g. trichostrongyle nematodes) which is often accompanied by the leakage of serum constituents and haemorrhage into the gut lumen (e.g. coccidians).

Tissue parasites may cause damage in the vicinity of the site of parasitization. In the case of *Wuchereria* the blockage and atrophy of the lymphatics leads to elephantiasis in a small proportion of infections. Other helminths such as *Ascaris* and *Fasciola* often cause marked, if transient, pathological effects during their migration to the site of parasitization. Passage of *Ascaris* through the lungs may produce symptoms akin to those of pneumonia, whilst the burrowing of *Fasciola* through the liver may cause death due to liver failure or secondary bacterial infection.

The various species of malaria and the African trypanosomes are the most serious protozoan parasites of the human bloodstream. Trypanosomes may invade the central nervous system a matter of weeks after infection giving rise to the characteristic shaking and sleeping symptoms, and death may result from a variety of associated conditions.

The most obvious symptoms of malaria are the regular bouts of fever which coincide with the synchronous release of merozoites from erythrocytes into the bloodstream. The malarial attack can last for weeks or months and untreated infections of *P. falciparum* are usually fatal. When erythrocytes rupture, the malarial pigment which they contain is dispersed through the body and taken up by phagocytes. It appears to be physiologically inert and is almost certainly not the factor which initiates the sequence of disease in the body. This function has been assigned to a so-far unidentified compound which depresses mitochondrial respiration in many different tissues. The chain reaction which is started involves the disruption of hormone balance and intense activity in the sympathetic nervous system. There is localized constriction of blood vessels and change in the pattern of blood flow to various organs. Fluids and proteins may escape from damaged blood vessels into tissues and the blood flow becomes sluggish or stops altogether. The result is a gradual deterioration of body function, the changes become irreversible, and the victim suffers physiological collapse and death.

Parasitism is a complex phenomenon whose investigation provides an enduring intellectual challenge to successive generations of parasitologists. It is a discipline worthy of study for its own sake but this brief account of some mechanisms of disease production is a fitting reminder of its wider importance. Protozoan and helminth parasites continue to exert a very real impact on the life of man and his domestic livestock.

References

ADAM, K. M. G., PAUL, J. and ZAMAN, V. (1971). *Medical and Veterinary Protozoology*. Churchill Livingstone, Edinburgh and London.

AIKAWA, M. and STERLING, C. R. (1974). *Intracellular Parasitic Protozoa*. Academic Press, New York and London.

BAKER, J. R. (1969). *Parasitic Protozoa*. Hutchinson, London.

BIRD, A. F. (1971). *The Structure of Nematodes*. Academic Press, London.

VAN DEN BOSSCHE, H. (Ed.) (1972). *Comparative Biochemistry of Parasites*. Academic Press, New York and London.

VON BRAND, T. (1973). *Biochemistry of Parasites*, 2nd edition. Academic Press, New York and London.

CANNING, E. U. and WRIGHT, C. A. (1972). *Behavioural Aspects of Parasite Transmission*. Academic Press, New York and London.

CHENG, T. C. (1973). *General Parasitology*. Academic Press, New York and London.

COHEN, S. and SADUN, E. H. (Eds) (1976). *Immunology of Parasitic Infections*. Blackwell, Oxford.

CROLL, N. A. (1970). *The Behaviour of Nematodes*. Edward Arnold, London.

CROLL, N. A. (1976). *The Organisation of Nematodes*. Academic Press, London.

CROLL, N. A. and MATTHEWS, B. E. (1977). *Biology of Nematodes*. Blackie, Glasgow and London.

ERASMUS, D. A. (1972). *The Biology of Trematodes*. Edward Arnold, London.

GARNHAM, P. C. C. (1966). *Malaria Parasites*. Blackwell, Oxford.

GUTTERIDGE, W. E. and COOMBS, G. H. (1977). *Biochemistry of Parasitic Protozoa*. Macmillan, London.

HAMMOND, D. M. and LONG, P. L. (1973). *The Coccidia*. Butterworths, London.

HOARE, C. A. (1972). *The Trypanosomes of Mammals*. Blackwell, Oxford.

JORDAN, P. and WEBBE, G. (1969). *Human Schistosomiasis*. Heinemann, London.

KENNEDY, C. R. (1975). *Ecological Animal Parasitology*. Blackwell, Oxford.

KENNEDY, C. R. (Ed.) (1976). *Ecological Aspects of Parasitology*. North-Holland, Amsterdam.

LEE, D. L. and ATKINSON, H. J. (1976). *Physiology of Nematodes*, 2nd edition. Macmillan, London.

LYONS, K. M. (1978). *The Biology of Helminth Parasites*. Studies in Biology no. 102. Edward Arnold, London.

MACDONALD, G. (1973). *Dynamics of Tropical Disease*. Oxford University Press, London.

MULLIGAN, H. W. (ed.) (1970). *The African Trypanomiases*. George Allen and Unwin, London.

NOBLE, E. R. and NOBLE, G. A. (1976). *Parasitology: The Biology of Animal Parasites*, 4th edition. Lea and Febiger, Philadelphia.

PETERS, W. and GILLES, H. M. (1977). *A Colour Atlas of Tropical Medicine and Parasitology*. Wolfe Medical Publications, London.

SASA, M. (1976). *Human Filariasis: A Global Survey of Epidemiology and Control*. University Park Press, Baltimore.

SCHMIDT, G. D. and ROBERTS, L. S. (1977). *Foundations of Parasitology*. C. V. Mosby, St Louis.

SMYTH, J. D. (1966). *The Physiology of Trematodes*. Oliver and Boyd, Edinburgh.

SMYTH, J. D. (1969). *The Physiology of Cestodes*. Oliver and Boyd, Edinburgh.

SMYTH, J. D. (1976). *Introduction to Animal Parasitology*, 2nd edition. Hodder and Stoughton, London.

SOLOMON, M. E. (1976). *Population Dynamics*, 2nd edition. Edward Arnold, London.

TAYLOR, A. E. R. and MULLER, R. (Eds) (1977). *Parasite Invasion*. Blackwell, Oxford.

WRIGHT, C. A. (1971). *Flukes and Snails*. George Allen and Unwin, London.